Contents

How To Use This Book

The ***Year 4 Targeting HASS Activity Book*** is filled with exciting stimulus materials to assist you in developing a greater understanding of the world you live in—its past and its present. The activities have been designed to encourage you to share opinions, think creatively, analyse ideas and information and respond in a variety of ways.

Each unit delivers content from the knowledge and understanding strand of the Australian Curriculum and supports the key inquiry questions in each subject area. The units also allow you to develop inquiry skills through:

Researching

Students identify and collect information, evidence and data from primary and secondary sources.

Questioning

Students develop questions about events, people, places, ideas, developments and issues to guide their investigations and satisfy their curiosity.

Analysing

Students explore information, evidence and data to identify and interpret features, patterns, trends and relationships, key points, facts and opinions and points of view.

Communicating

Students present ideas, findings, viewpoints, judgements and conclusions in digital and non-digital forms for different audiences and purposes.

Evaluating and Reflecting

Students propose explanations for events, developments and issues, draw evidence-based conclusions and use criteria to make informed decisions and judgements.

This book is organised into the Year 4 curriculum areas, as follows:

- **Units 1–14: HISTORY**
- **Units 15–26: GEOGRAPHY**
- **Units 27–32: CIVICS AND CITIZENSHIP**

TARGETING HASS 4 © PASCAL PRESS ISBN: 9781925726053

The *Targeting HASS Activity Book* series is designed to be a flexible learning resource. The units do not have to be completed in numerical order. Your teacher may direct you to complete units and activities from one learning area before moving on to a different learning focus. In this way, the *Targeting HASS* series will complement any school's scope and sequence.

The seven **Australian Curriculum's General Capabilities** listed below are embedded in the units of this book:

- Literacy
- Numeracy
- Information and Communication Technology (ICT) Capability
- Critical and Creative Thinking
- Personal and Social Capability
- Ethical Understanding
- Intercultural Understanding

A number of the activities and questions are open-ended. This allows for student challenge and differentiation. The questions are written to encourage you to think deeply about topics and provide extended responses. For some open-ended questions, answers are not provided. These questions are best graded through peers marking each other's work, or through student–teacher conferences. There are brief sample responses to guide the assessment of your work for some open-ended questions.

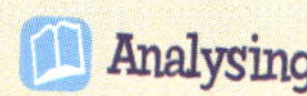

Analysing

What is the purpose of the image showing cleared forest in Borneo? How does it impact the reader?

Evaluating and Reflecting

Can you think of a time when rationing might be needed? What would be rationed and why would its use need to be carefully managed?

The final section features eight **assessment** activities:

- 3 History
- 3 Geography
- 2 Civics & Citizenship.

Each assessment activity relates to a key inquiry question. Make sure you have completed the units for that inquiry question before tackling the assessment.

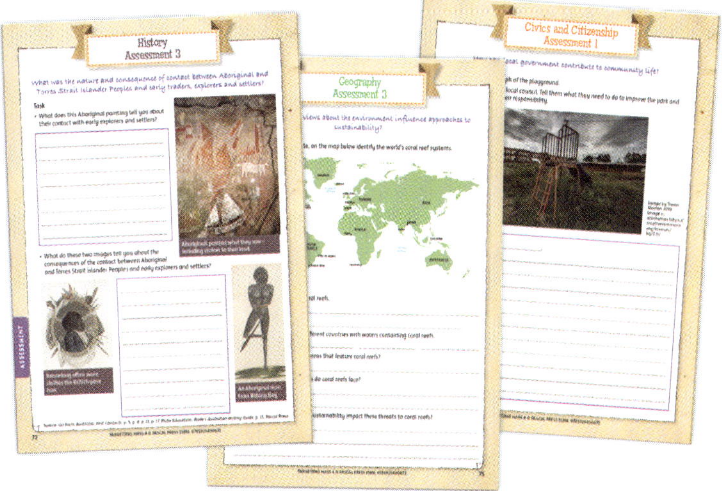

Additional inquiry skills covered by each unit

		ACARA CODE	WA CODE	VIC CODE
Questioning	Pose questions for investigations	ACHASS1073	WAHASS27	
Researching	Research, locate and collect data	ACHASS1074	WAHASS28	VCGGC074 VCHHC067
	Record, sort and represent data	ACHASS1075	WAHASS29	VCGGC075
	Sequence information	ACHASS1076		VCHHC066
Analysing	Examine information and distinguish facts from opinions	ACHASS1077	WAHASS33	
	Interpret data	ACHASS1078	WAHASS32	VCGGC076
Communicating	Present ideas, findings and conclusions	ACHASS1082	WAHASS37	
Evaluating and Reflecting	Draw simple conclusions based on analysis of information	ACHASS1079	WAHASS35	
	Interact with others to share points of view	ACHASS1080	WAHASS36	
	Reflect on learning and consider possible effects of proposed actions	ACHASS1081	WAHASS39	

Curriculum correlations

ACARA CODE: DESCRIPTION – HISTORY
ACHASSK083 The diversity of Australia's first peoples and the long and continuous connection of Aboriginal and Torres Strait Islander Peoples to Country/Place
ACHASSK084 The journey(s) of at least one world navigator, explorer or trader up to the late 18th Century, including their contacts with other societies and any impacts
ACHASSK085 Stories of the First Fleet, including reasons for the journey, who travelled to Australia, and their experiences following arrival
ACHASSK086 The nature of contact between Aboriginal and Torres Strait Islander Peoples and others, for example, the Macassans and the Europeans, and the effects of these interactions on, for example, people and environments
ACARA CODE: DESCRIPTION – GEOGRAPHY
ACHASSK087 The main characteristics of the continents of Africa and South America and the location of their major countries in relation to Australia
ACHASSK088 The importance of environments, including natural vegetation, to animals and people
ACHASSK089 The custodial responsibility Aboriginal and Torres Strait Islander Peoples have for Country/Place, and how this influences views about sustainability
ACHASSK090 The use and management of natural resources and waste and the different views on how to do this sustainably
ACARA CODE: DESCRIPTION – CIVICS AND CITIZENSHIP
ACHASSK091 The role of local government and the decisions it makes on behalf of the community
ACHASSK092 The differences between 'rules' and 'laws', why laws are important and how they affect the lives of people, including experiences of Aboriginal and Torres Strait Islander Peoples
ACHASSK093 The different cultural, religious and/or social groups to which they and others in the community belong
NSW HISTORY STAGE 2 CURRICULUM OUTCOMES
HT2-1 identifies celebrations and commemorations of significance in Australia and the world
HT2-2 describes and explains how significant individuals, groups and events contributed to changes in the local community over time
HT2-3 describes people, events and actions related to world exploration and its effects
HT2-4 describes and explains effects of British colonisation in Australia
HT2-5 applies skills of historical inquiry and communication
NSW GEOGRAPHY STAGE 2 CURRICULUM OUTCOMES
GE2-1 examines features and characteristics of places and environments
GE2-2 describes the ways people, places and environments interact
GE2-3 examines differing perceptions about the management of places and environments
GE2-4 acquires and communicates geographical information using geographical tools for inquiry
VICTORIA HISTORY LEVELS 3 & 4 CURRICULUM OUTCOMES
VCHHK072 The significance of Country and Place to Aboriginal and Torres Strait Islander peoples who belong to a local area
VCHHK074 The role that people of diverse backgrounds have played in the development and character of the local community and/or other societies
VCHHK078 The diversity and longevity of Australia's first peoples and the significant ways Aboriginal and Torres Strait Islander peoples are connected to Country and Place (land, sea, waterways and skies) and the effects on their daily lives
VCHHK079 The journey(s) of a significant world navigator, explorer or trader up to the late eighteenth century, including their contacts with and effects on other societies
VCHHK080 Stories of the First Fleet, including causes and reasons for the journey, who travelled to Australia, and their experiences and perspectives following arrival
VCHHK081 The nature of contact between Aboriginal and Torres Strait Islander peoples and others, for example, the Macassans and the Europeans, and the effects of these interactions
VICTORIA GEOGRAPHY LEVELS 3 & 4 CURRICULUM OUTCOMES
VCGGK077 Location of major countries of Africa and South America in relation to Australia and their major characteristics including the types of vegetation and native animals in at least two countries for both continents
VCGGK078 Location of Australia's neighbouring countries and the diverse characteristics of their places
VCGGK080 The many Countries/Places of Aboriginal and Torres Strait Islander peoples throughout Australia, and the custodial responsibility they have for Country/Place, and how this influences views about sustainability
VCGGK081 Main climates of the world and the similarities and differences between the climates of different places
VCGGK082 Types of natural vegetation and the significance of vegetation to the environment, the importance of environments to animals and people, and different views on how they can be protected; the use and management of natural resources and waste, and different views on how to do this sustainably
VCGGK083 Similarities and differences in individuals' and groups' feelings and perceptions about places, and how they influence views about the protection of these places
VICTORIA CIVICS & CITIZENSHIP LEVELS 3 & 4 CURRICULUM OUTCOMES
VCCCG001 Identify features of government and law and describe key democratic values
VCCCG002 Identify how and why decisions are made democratically in communities
VCCCG003 Explain the roles of local government and some familiar services provided at the local level
VCCCL004 Explain how and why people make rules
VCCCL005 Investigate why and how people participate within communities and cultural and social groups

TARGETING HASS 4 © PASCAL PRESS ISBN: 9781925726053

History														Geography												Civics and Citizenship						History Assessment			Geography Assessment			Civics & Citizenship Assessment	
1	2	3	4	5	6	7	8	9	10	11	12	13	14	15	16	17	18	19	20	21	22	23	24	25	26	27	28	29	30	31	32	1	2	3	1	2	3	1	2
				✔	✔	✔																																	
✔	✔	✔	✔																													✔							
							✔	✔	✔	✔																							✔						
											✔	✔	✔																					✔					
														✔	✔	✔	✔																		✔				
																		✔	✔	✔																			
																					✔	✔															✔		
																							✔	✔	✔											✔			
																										✔	✔											✔	
																												✔	✔										✔
																														✔	✔								
																														✔	✔								
				✔	✔	✔																																	
✔	✔	✔	✔				✔	✔	✔	✔	✔	✔	✔																			✔							
							✔	✔	✔	✔	✔	✔	✔																				✔	✔					
✔	✔	✔	✔	✔	✔	✔	✔	✔	✔	✔	✔	✔	✔																			✔	✔	✔					
														✔	✔	✔	✔	✔	✔	✔	✔	✔	✔	✔	✔										✔	✔	✔		
																	✔	✔	✔	✔	✔	✔	✔	✔	✔											✔	✔		
																	✔	✔	✔	✔	✔	✔	✔	✔	✔										✔	✔			
														✔	✔	✔	✔	✔	✔	✔	✔	✔	✔	✔	✔										✔		✔		
					✔							✔	✔																	✔				✔					
																															✔								
				✔	✔	✔																																	
✔	✔	✔	✔																													✔							
							✔	✔	✔	✔																							✔						
											✔	✔	✔																										
														✔	✔	✔															✔				✔		✔		
														✔																									
																						✔																	
														✔																									
																	✔	✔	✔	✔			✔	✔	✔											✔	✔		
																		✔	✔		✔									✔									
																										✔	✔											✔	
																										✔	✔											✔	
																																						✔	
																												✔	✔										✔
																												✔	✔										✔

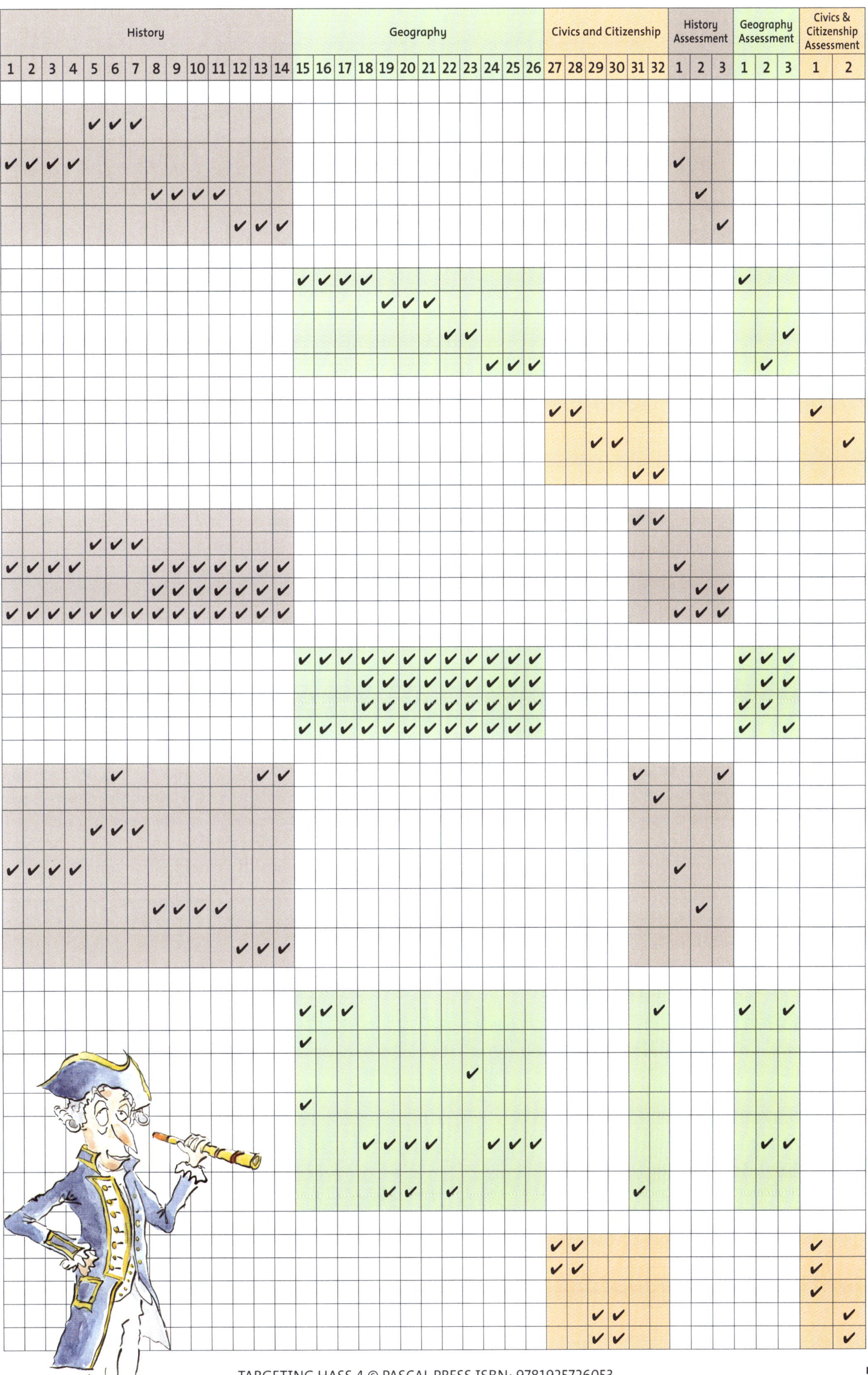

HISTORY

A Wider World

Between the 15th and 17th centuries, European nations were eager to explore the world and take the first steps towards globalisation.

The period became known as The Age of Exploration. But with no modern radios, satellite navigation or Global Positioning Systems (GPS), sailing on a small wooden ship across uncharted waters during a time when people truly believed that they would sail off the edge of the world if they went too far was very frightening. Many also believed that beneath the oceans lived sea monsters such as the kraken, that were capable of swallowing ships whole.

So why did they? Put simply, it was for 'God, Glory and Gold'. Kings and queens paid for such explorations because they wanted to grow their empires. They wanted personal fame and to bring honour to their nations. They wanted to find new trade routes and claim the wealth that could be acquired from other parts of the world. This was not only the gold that was stolen from people who could not defend themselves, but also by trading for things such as silk and spices which were luxuries at the time.

Global trade began to develop between the Old world (Europe, Asia and Africa) and the New World (the Americas and later on Australia). With this came a situation known as the Columbian exchange, where there was a significant transfer of plants, animals, food and people (including slaves) between the Eastern and Western hemispheres. But as the European nations competed to discover new worlds, their explorers also introduced diseases to the people they met. Waves of smallpox, diphtheria and measles killed millions of indigenous people.

The Europeans also believed that they were superior to the people they 'discovered' on their journeys. They saw it as their duty to teach their language, religion and way of life to the natives who they called 'heathens'. Explorers frequently took priests and monks with them on their journeys, to introduce Christianity to whomever they found. As a result, Christianity became the world's largest religion.

The Aztec capital city of Tenochtitlan fell to the Spanish in 1521. Its ruins are in the modern capital of Mexico, Mexico City.

Source: Go Facts Australia: *Explorers*, p. 4, p. 5, p. 7, Blake Education; *Blake's Australian History Guide*, p. 6, Pascal Press

TARGETING HASS 4 © PASCAL PRESS ISBN: 9781925726053

Research

Many European nations had established colonies in the East (Asia) and were trading with them by going overland.

Discover why finding a sea route was so important.

____

____

____

____

____

____

____

Questioning

What do the following words mean?

- heathen: ______________________________
____
- globalisation: ______________________________
____
- kraken: ______________________________
____
- diphtheria: ______________________________
____
- uncharted: ______________________________
____

Analysing

Compare the risks faced by early explorers travelling to new worlds with the risks faced by space explorers today.

Risks faced by early sea explorers	Risks faced by space explorers

Communicating

Would you like to be an explorer, travelling to new worlds and experiencing the unknown?

Prepare a short speech to explain your point of view.

Evaluating and Reflecting

Do the benefits of exploration – the gaining of new knowledge and the sharing of goods in trade – outweigh the negative consequences, such as the spread of disease and loss of traditional land ownership? Explain why.

____

____

____

____

UNIT 2

HISTORY

Trade during the Age of Discovery

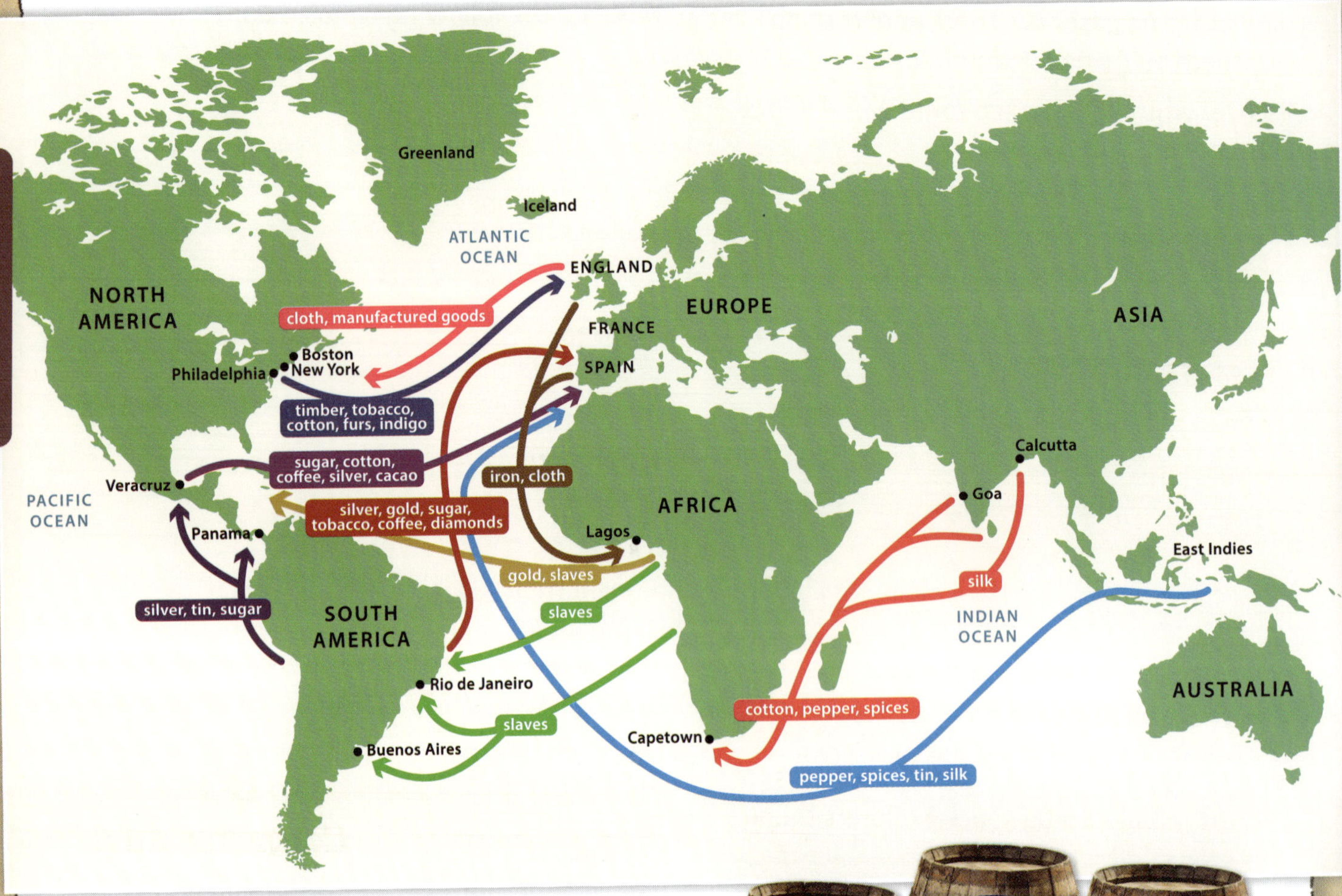

Greenland
Iceland
ATLANTIC OCEAN
ENGLAND
FRANCE
SPAIN
EUROPE
ASIA
NORTH AMERICA
cloth, manufactured goods
Boston
New York
Philadelphia
timber, tobacco, cotton, furs, indigo
sugar, cotton, coffee, silver, cacao
iron, cloth
Calcutta
Goa
PACIFIC OCEAN
Veracruz
silver, gold, sugar, tobacco, coffee, diamonds
AFRICA
Lagos
Panama
gold, slaves
East Indies
silk
silver, tin, sugar
SOUTH AMERICA
slaves
INDIAN OCEAN
Rio de Janeiro
AUSTRALIA
cotton, pepper, spices
slaves
Capetown
Buenos Aires
pepper, spices, tin, silk

Some of the main aims of exploration in the past were to discover new trade routes, new goods to trade and new people to trade with. When explorers from Europe sailed to new lands, they discovered all kinds of plants and animals they had never seen before. They took samples of these back to Europe, Africa and Asia. Goods such as sugar, cotton, coffee, gold, silver, tobacco, spices and silk were some of the items that were traded.

Trade also changed the eating habits of people all over the world. For example, Europeans had never tried tomatoes until explorers brought them back from the Americas. Other foods brought from the Americas included potatoes, beans, squash, avocados, pineapples, tobacco, chilli peppers and chocolate.

In return, European explorers took animals to the Americas — such as cows, goats, sheep and pigs. They also brought new ideas and new ways of doing things. Through exploration, an exchange of knowledge and skills started to take place.

Unfortunately, another type of trade was occurring at the time — the slave trade. This was the exchange of people. They were abducted from Africa and sold to work on plantations in Europe and the Americas.

Source: Australian History Centres: *Middle Primary*, p.47, Blake Education

TARGETING HASS 4 © PASCAL PRESS ISBN: 9781925726053

Research

Select one trade good shown on the map.
Is it still produced in its country of origin?
How much is produced and how is it traded today?

Questioning

When you look at the map, what questions do you ask yourself?
Write at least three questions.

1

2

3

Analysing

Explain why there are no lines on the map showing trade with Australia.

What do the trading centres marked in Europe, Africa and America all have in common?

Communicating

What are the benefits of trading goods between countries? Discuss this in a small group and develop a mind map to show your ideas.

Evaluating and Reflecting

Do we still have to trade for all the goods listed on the map?
Explain how the trade in some of these goods has changed. Use at least two specific examples.

Exploration in the Past

CHRISTOPHER COLUMBUS (1451–1506)

Columbus was born in Genoa, Italy, but it was King Ferdinand and Queen Isabella of Spain who financed his famous voyages. Portugal had gained fame and power from their successes at sea, and Spain wanted the same glory. In 1492, Columbus led three ships (*Niña*, *Pinta* and *Santa Maria*) across the Atlantic Ocean from Spain to what is now the West Indies.

Map of Columbus's first voyage

Columbus's great achievement was not only that he 'discovered' America, but that he proved to everyone that the world wasn't flat and a ship wouldn't fall over the edge if it sailed too far from the coast. Of course, many others had thought this for a long time, but actually demonstrating it was not only a frightening experience, but also a notable one. However, Columbus thought he had sailed halfway around the world and reached India. He truly had no idea that he had discovered an entirely new continent unknown to the Europeans.

Columbus monument, Barcelona

FERDINAND MAGELLAN (1480–1521)

Magellan was born in Sabrosa in northern Portugal, and, like Christopher Columbus before him, he served Spain (under King Ferdinand's successor, Charles V). After Columbus reached the Americas, it was soon realised that it was not the westward route to the 'Spice Islands' (part of modern Indonesia). The Europeans wanted a better route to the Spice Islands than sailing around the Cape of Good Hope.

Magellan's expedition of 1519 to 1522 was possibly one of the most daring and risky journeys ever taken. It became the first expedition to sail from the Atlantic Ocean into the Pacific Ocean, and completed the first circumnavigation of the globe. Magellan himself did not complete the voyage, as he was killed by natives in the Philippines in 1521.

Magellan's Cross, Philippines

Look into the history of world circumnavigation. People also discovered new lands and regions through flight. Find out when the world was first circled by aircraft. How quickly can it be done now by using spacecraft?

The trade that Europeans carried out with the East was in goods such as silk and spices. Why do you think these would have been so profitable and what other goods were traded?

Source: *Blake's Australian History Guide*, p. 6, Pascal Press

TARGETING HASS 4 © PASCAL PRESS ISBN: 9781925726053

Research

Choose either Christopher Columbus or Ferdinand Magellan.

Research the impact their exploration had on at least one indigenous group of people from the lands they visited.

Questioning

1 What myth did Columbus disprove?

2 What mistake did Columbus make?

3 What was the purpose of Magellan's journey?

4 What were two important achievements of Magellan's expedition?

Analysing

Did Columbus really 'discover' America? Explain your answer.

Communicating

On a map of the world, plot the journeys of Christopher Columbus and Ferdinand Magellan. Show where they went and the ports they stopped at.

Evaluating and Reflecting

Which expedition was more important – Columbus's or Magellan's? Explain why.

UNIT 4

Trials and Travels: Abel Tasman

HISTORY

Undaunted by earlier failed missions, the Governor-General of Batavia, Anthony van Dieman, put his faith in 39-year-old Captain Abel Jansz Tasman to help the Dutch add to their territory in the Pacific. In the *Heemskerck* and *Zeehaen*, Tasman discovered Van Dieman's Land (Tasmania) before being blown east to New Zealand.

Abel Tasman, depicted with his wife and daughter in 1637, in a painting believed to be by Jacob Gerritsz Cuyp.

TASMAN'S VOYAGE, while it did not accomplish what it set out to do (which was to ascertain whether the land to the south was joined to Antarctica or was an island, and to find "uncommonly large profit"), was nevertheless successful in part.

ON 3 DECEMBER 1642, Tasman claimed Van Dieman's Land for the Dutch. Strangely, he did not attempt to circumnavigate the island he found; nor did he travel around New Zealand later in the journey. Tasman was impressed with Van Dieman's Land, writing:

> *Whoever perfectly discovers and settles it will become infallibly possessed of Territories as Rich, as fruitful, & as capable of improvement, as any that have been hitherto found out, either in the East Indies or the West.*

TASMAN WORKED FOR the powerful Dutch East India Company, which made a fortune out of trading spices and wood from the East Indies (now Indonesia), and other luxury items such as silk, coffee, porcelain, gold and copper from around the world. In spite of Tasman's enthusiastic report, the company was disappointed. In 1644, van Dieman sent him on a second mission to figure out whether New Guinea, New Holland, Van Dieman's Land and the new land (New Zealand) were linked. On this voyage, Tasman charted much of the north Australian coast and all of the Gulf of Carpentaria before returning to Batavia.

the FACTS!

TASMAN'S JOURNEY was made for the dual purposes of discovery and trade. If Tasman found inhabitants, van Dieman instructed him to, *"parley with the rulers and subjects, letting them know you have landed there for the sake of commerce"*.

THE STRANGE LIST of provisions on Tasman's vessels included 19 pounds of elephant teeth, 200 small Chinese wooden combs, a large brass basin and 2 packets of tinsel!

TASMAN AND HIS CREW heard noises when they went ashore but did not see any Aborigines. They did see footholds carved into a tree – the height of the footholds made them think the natives in that area must have been giants.

LATER IN HIS CAREER, Tasman became a buccaneer and tried to make his living from plundering Spanish ships.

Abel Tasman's ships were the *Heemskerck* (Dutch for "home church"), which carried 60 men, and the *Zeehaen* (the "sea cock" or "rooster"), which had a crew of 50.

Source: Amazing Facts about Australia: *Early Explorers*, p. 15, Steve Parish

TARGETING HASS 4 © PASCAL PRESS ISBN: 9781925726053

Research

Abel Tasman had two major voyages – one in 1642 and the other in 1644. Select one to research further. Write four things about this voyage that you didn't already know.

1 ______________________________

____

____

2 ______________________________

____

____

3 ______________________________

____

____

4 ______________________________

____

____

Questioning

You are going to quiz a classmate on the information about Abel Tasman and his voyages. Write four questions to ask them.

1 ______________________________

____

____

2 ______________________________

____

____

3 ______________________________

____

____

4 ______________________________

____

____

Analysing

Why do you think Abel Tasman had such a strange list of provisions on his vessels?

____

____

____

Communicating

Create a collage of images to support Tasman's enthusiastic report about the qualities of Tasmania.

Evaluating and Reflecting

Should the Dutch East India Company have been disappointed with Abel Tasman and his expeditions? Explain your answer.

____

____

____

____

Torres Strait Islander Peoples

The origin, history and ways of life of Torres Strait Islander peoples are very different to those of Aboriginal peoples who live on mainland Australia. This is because Torres Strait Islander peoples have Melanesian origins – they are not descendants of mainland Aboriginal peoples migrating to live on the islands of the Torres Strait.

A map of Australia and Papua New Guinea. The Torres Strait region is shown by the circle and arrow.

As far back as 80,000 years ago, there was a land bridge that joined New Guinea to Australia. This landmass was called Sahul and made it possible for people to walk from Southern New Guinea to what is now the Cape York Peninsula in Queensland. It is believed that the first people migrated to the Torres Strait about 70,000 years ago. At the end of the last ice age, about 12,000 years ago, the sea levels rose and the land bridge was covered with water. This formed the Torres Strait, which connects the Arafura and Coral Seas. Archaeologists have found evidence of people living on the Torres Strait that dates back to 2500 years ago.

The Torres Strait is made up of more than 100 small islands. Today, people live in communities on just 17 of these islands. In the past, many more islands were inhabited. People built outrigger canoes to travel between islands to hunt, fish and collect food.

The outrigger canoe got its name because long weights called 'outriggers' were used on one or both sides to help the canoe stay upright. The materials used to make the canoes came from natural resources, such as pandanus palm leaves for the sails. People traded items like ornaments and conus shells for materials they didn't have. The wood for the hull would have had to come from another island, like Papua New Guinea.

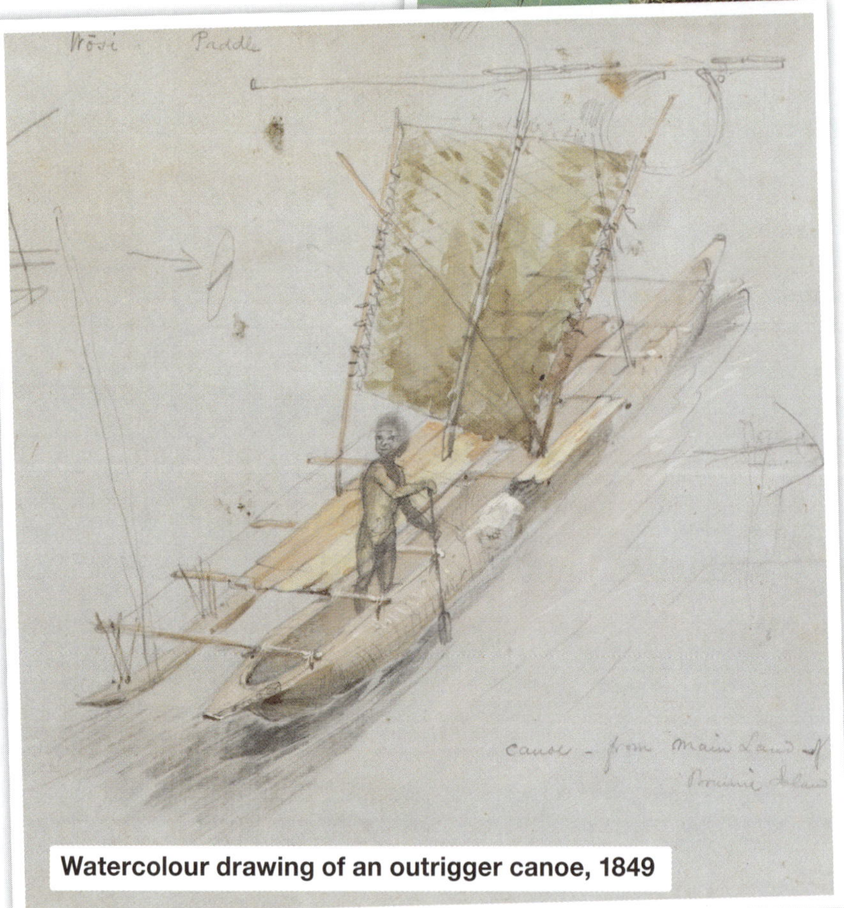

Watercolour drawing of an outrigger canoe, 1849

Source: Australian History Centres: *Middle Primary*, p. 57, Blake Education

TARGETING HASS 4 © PASCAL PRESS ISBN: 9781925726053

Research

The people of the Torres Strait Islands have their own flag. Find out what the different symbols on the flag represent.

Questioning

1 Why is the culture and way of life of the mainland Aboriginal peoples so different to those of Torres Strait Islander peoples?

2 What was Sahul?

3 What are 'outriggers' on a canoe?

4 What bodies of water are either side of the Torres Strait?

Analysing

What would have been some of the similarities and differences between the lives of traditional Aboriginal peoples and the lives of Torres Strait Islanders?

Similarities	Differences

Communicating

Using repurposed materials, create a model of a dugout canoe.

Explain your design process and discuss any problems you had to solve.

Evaluating and Reflecting

What was the impact of rising sea levels at the end of the ice age? What might be different now if the bridge had not been covered?

First peoples: Tasmania and Western Australia

TASMANIA

Tasmanian Aboriginal peoples, the Palawa, have lived in Tasmania for 40 000 years. At that time, Tasmania was connected to the mainland by a land bridge that people used to migrate to the southern peninsula. Around 13 000 years ago, melting ice caps caused the sea levels to rise and flood the land bridge, creating the island of Tasmania and cutting off the Palawa from the rest of Australia. The body of water that covers the area where the land bridge once existed is now known as Bass Strait.

Artist's depiction of Tasmanian Aboriginal peoples, ca. 1822

The Palawa were primarily hunter-gatherers. From the hinterland to the coast, the Palawa enjoyed a range of abundant food sources, which varied throughout the seasons. On land they had access to wallaby and possum, roots and berries, and fruits and vegetables. They used canoes to travel to surrounding islands, where they would hunt muttonbirds and seals. Along the coastline, ancient middens are evidence of a diet that included abalone, oysters, mussels and shellfish.

The Palawa people were cut off from Aboriginal peoples on the mainland for several thousand years. Investigate and discuss how this caused their culture and traditions to evolve differently to those of the Noongar people.

WESTERN AUSTRALIA

The south-west of Western Australia has been occupied by the Noongar people for the last 45 000 years. Like Aboriginal peoples in other parts of the country, the Noongar placed much importance on their relationship with the environment. They were hunter-gatherers who used their knowledge of the seasons when finding and collecting food to ensure the survival of these plants and animals for the years to come.

Many early Aboriginal peoples were hunter-gatherers, and their relationship with the environment was very important to them. Read about the six Noongar seasons at www.bom.gov.au/iwk/nyoongar/ and draw a poster depicting each season and its plants and animals.

These food sources varied according to the parts of the south-west in which the different groups of Noongar people lived. Those who occupied the Perth area had access to the ocean, the Swan River and freshwater lakes. In the south and east, food came from the forests. Men hunted daily for larger animals like kangaroo, while women and children gathered smaller animals, such as echidnas and lizards. Along the southern coast, people caught fish and hunted turtle.

Rottnest Island

Source: *Blake's Australian History Guide*, p. 20, Pascal Press

Research

Investigate the six Noongar seasons.
What are they and how are they identified?

1 ______________________________

____

2 ______________________________

____

3 ______________________________

____

4 ______________________________

____

5 ______________________________

____

6 ______________________________

____

Questioning

What questions do you now have about the Aboriginal people of Tasmania or Western Australia?

Write at least three questions. Can you find out the answers to your questions?

____

____

____

____

____

____

____

____

Analysing

Look at the artist's depiction of Tasmanian Aboriginal peoples ca. 1822. What does it tell you about these people and their way of life?

____

____

____

Communicating

Aboriginal people used art to share stories of the Dreaming and their connection with country.

Create your own artwork to share with others, showing your connection to a special place or belief.

Evaluating and Reflecting

How did the separation of Tasmania from the Australian mainland around 13 000 years ago impact the diversity of Australia's first peoples? What this impact positive or negative? Explain.

____

____

____

____

HISTORY

Aboriginal Shell Middens

A shell midden is a place where leftover items from eating food like shellfish have piled up over a long time. The main types of shells found in middens are from oysters and mussels. Other items found in shell middens include animal bones from birds and fish, and charcoal from making campfires. Some middens even have tools made from stone, shell or bone. Shell middens are made up of layers that build up over time. These different layers show us what people ate at the different times the shell midden was used as a place to throw things away.

Shell midden layers, Moreton Bay, Queensland

Shell midden erosion, Steele's Island, Tasmania

Shell midden, West Point, Tasmania

Sydney Rock Oysters

There are shell middens all over Australia. These were created by Aboriginal peoples, and the shell middens tell us what Aboriginal peoples ate and how they lived thousands of years ago. Shell middens are usually located in coastal areas where people were able to find shellfish easily. They can be in places like sandy beaches and sand dunes, and along the banks of rivers and lakes. They can be as short as 1 metre, and as long as 1000 metres (1 kilometre).

As sea levels slowly rise, the waves wash away sand on the beach that the sea couldn't reach before. This is known as 'erosion'. Some shell midden sites on the coasts are being slowly eroded (washed away). They could be lost forever in the next 30 to 50 years.

Source: Australian Geography Centres: *Lower Primary*, p. 17, Blake Education

TARGETING HASS 4 © PASCAL PRESS ISBN: 9781925726053

Research

What is erosion? How does it impact important historical sites like shell middens?

Questioning

What questions about traditional Aboriginal ways of life could be answered by examining shell middens?

Write at least three questions.

Analysing

How would midden sites vary from one place to another across Australia?

Communicating

Explain why protecting historical sites such as shell middens is important.

You may do this by presenting a one-minute speech or creating a poster with text and images.

Evaluating and Reflecting

In the future, what could archaeologists learn about our society by examining our leftover items (rubbish)?

UNIT 8

The British are Coming

Hard Times

In the eighteenth and nineteenth centuries, life in Britain was very hard for the poor. The industrial revolution meant that many people were out of work, because goods that were once made by hand could now be produced quickly and cheaply by machines in factories. Unlike today, there was no unemployment assistance like welfare from the government, and there were few laws to protect working people from greedy and cruel bosses. Those who fell on hard times had to struggle as best they could – the homeless often turning to begging or pickpocketing as a result. If convicted of these crimes, punishment was harsh. Stealing animals or other goods worth more than five shillings often meant a death sentence.

Prison Ships

Prisons were overcrowded and terrible places. To make more room for prisoners, old boats known as hulks were used as prison ships. Hulks were even more terrible than prisons on land, and a third of the prisoners kept on them died due to the spread of disease. The use of hulks to solve the overcrowding issue was neither a humane nor long-term solution. Britain had been transporting its excess prisoners to its North American colonies, but America's independence in 1783 brought an end to this practice. Instead, Britain turned to its land in New South Wales, which had been claimed for Britain by Captain James Cook in 1770.

Criminal Children

The people of nineteenth-century Britain were very worried about the growing crime rate. They believed that treating criminals cruelly helped to reform them and making a spectacle of public hangings helped to discourage others from committing crimes. This attitude also applied to children, who were sent to adult prisons, with some as young as 12 facing the hangman's noose. Eventually, reformatory schools were established for young offenders but living conditions at the schools were tough. Orphaned children often found themselves huddling together in dark alleys and lanes (known as rookeries) at night.

The Proposal

The overcrowding in prisons and onboard floating hulks was not the only supporting reason for establishing New South Wales as a penal colony. There were raw materials to be exploited and the colony could be used as an entry point for the economic opportunities of the surrounding region. The continent was also of strategic importance in the struggle against the Portuguese, Dutch and French. The British were sure the convicts would be able to grow food and support themselves, based on the evidence of naturalist Joseph Banks who had been on Captain Cook's ship *Endeavour* when it arrived on Australian shores in 1770.

Source: *Blake's Australian History Guide*, pp. 21-22, Pascal Press; Amazing Facts about Australia: *Early Settlers*, p. 5, Steve Parish

TARGETING HASS 4 © PASCAL PRESS ISBN: 9781925726053

Research

What was life like for children who had been transported to the colony of New South Wales for petty crimes?

Find out more about their stories by visiting the Sydney Living Museums website, or other information sources.

Questioning

At nine years of age, John Hudson was the youngest convict on the First Fleet.

If you could talk to John when he arrived in Sydney Cove, what would you ask? Write at least four questions.

Analysing

Compare and contrast life for children in Britain in the 18th and 19th centuries, and your life today.

Life in the 18th and 19th centuries	My life today

Communicating

Learn the song *Botany Bay*.

Create your own actions to match the lyrics, and present your performance to the class.

Evaluating and Reflecting

Do you think that harsh penalties, such as death sentences for stealing food, were effective in stopping people from breaking the law? Explain your answer.

HISTORY

Arrival of the First Fleet

The eleven ships of the First Fleet, carrying more than 1400 people, left Portsmouth in England on 13 May 1787. They arrived at Botany Bay 18–20 December 1787; however, Captain Arthur Phillip found the site unsuitable and moved to Port Jackson (now better known as Sydney Harbour) to found the new colony.

John Alcott's depiction of ships of the First Fleet anchored in Sydney Cove in 1788.

ON SATURDAY 26 JANUARY (now celebrated annually as Australia Day) the officers of the First Fleet raised the Union Jack, toasted the health of the royal family and the new colony, fired their muskets and cheered the founding of tho colony.

BUILDING A NEW LIFE

The first task was to establish a settlement. There was no suitable accommodation, so convicts and officers slept in tents until more permanent dwellings could be built. Male convicts were ordered to start constructing buildings soon after they landed, with some of the earliest projects being a hospital, wharves, stores, a church, soldiers' barracks and homes for the officers. The only convicts with any kind of experience were one brickmaker, five bricklayers, three plasterers, three carpenters and a stonemason. The rest were inexperienced for the task and many of the initial buildings collapsed. Some of the first flimsy huts were hastily constructed out of wattle-and-daub, which did not withstand the heavy summer storms. Buildings to house the convicts were last on the list and as a result some convict boys slept in a hollow tree for months while huts were being built!

To keep male and female convicts separated, female felons remained on the ships until Wednesday 6 February 1788. The following day, Governor Phillip addressed the convicts and told them that anyone attempting to escape would be shot immediately. Anyone attempting to steal food or animals would also face severe punishment. Those who refused to work would be denied food. "Many of you are innate villains," he noted. Female convicts were set to work crushing oyster shells to make lime for cement.

the FACTS!

WHEN THE FIRST FLEET sailed into Port Jackson, Surgeon White wrote: *"Port Jackson I believe to be, without exception, the finest and most extensive harbour in the universe …"*

LIMITED personal possessions were allowed onboard the First Fleet, but some of the affluent "gentlemen convicts" (usually convicted for forgery or fraud) may have carried a few books, a writing set or a piece of jewellery. Most left with just the shirts on their backs.

THE SUPPLY SHIPS of the First Fleet carried goods to help establish the colony, such as 14 000 shovels, 40 wheelbarrows, 8000 fish hooks and 5448 panes of glass.

A FEMALE CONVICT named Mary Bolton gave birth to the colony's first baby, Joshua, on one of the transport ships as the First Fleet entered Botany Bay. Babies of convicts were free-born and were labelled as BC on the government "muster" (or census) which stood for Born in the Colony.

Source: Amazing Facts about Australia: *Early Settlers*, p. 8, Steve Parish

Research

How would you describe the journey of the First Fleet? What route did it travel? What difficulties did they encounter?

Questioning

1 What impressed Surgeon White about the harbour at Port Jackson?

2 Why were babies labelled 'BC' on the government census?

3 Why did many of the first buildings in the colony collapse?

4 What action would result in a convict being refused food?

Analysing

What were the main factors that made life difficult in the new colony?

Communicating

Imagine you were starting a new colony of students in a part of the schoolground. Create a list of rules for everyone to follow.

Evaluating and Reflecting

Was it fair to build the convict houses last? Why do you think this?

A Colony of Convicts

Eleven ships carrying more than 1400 people, including officials, passengers and crew, left Portsmouth for the new land on 13 May 1787. Eight months later, the weary seafarers and convicts had their first glimpse of their new home.

The First Fleet's flagship *HMS Sirius*

GOVERNOR ARTHUR PHILLIP quickly dismissed Botany Bay as a suitable place for settlement. Instead travelling north, he entered Sydney Heads and saw the "finest harbour in the world" – Port Jackson, now better known as Sydney Harbour. He examined the many coves and settled on one "that had the best spring of fresh water", naming it Sydney Cove after Viscount Sydney, the then Home Secretary of the British Government. Phillip then returned to the ships and had them all moved to Sydney Cove.

PHILLIP AND HIS SORRY BAND of prisoners, officers and marines were the first European explorers of the hinterland of Australia's east coast. But their first tasks were to establish buildings, storehouses, roads, vegetable gardens and other hallmarks of settlement. The convicts and marines were ill-equipped and inexperienced for this task, but were put to work immediately, clearing bush and logging timber.

Artist's impression of convicts in Britain being rowed out to the First Fleet's ships.

WITHIN FIVE WEEKS, Phillip began to undertake another task that had been entrusted to him – further exploration around the colony. In a long boat, he visited Broken Bay and discovered Brisbane Water. Rowing west of the settlement he found the mouth of the Hawkesbury River and later remarked on the "finest piece of water I ever saw" – he named it Pitt Water, after William Pitt, the prime minister of England, but it is now known as Pittwater. By April, he had travelled as far as Lake Narrabean and noted "a very fine view of the mountains inland". These were the Blue Mountains, which surround the Sydney region and would later prove a frustrating obstacle to further exploration.

A WORTHY LEADER

Governor Phillip endeavoured to ensure fairness for all in the struggling colony. He was particularly mindful of the Aborigines:

> *I shall think it a great point gained if I can proceed … without having any dispute with the natives; a few of them I shall endeavour to settle near us and whom I mean to furnish with everything I can tend to civilise them, and give them a high opinion of their new guests.*

Phillip was the first to use Aboriginal guides and befriended many Aborigines over his time as governor. When he became ill after being speared by a hostile Aborigine at Manly, he returned to England, taking with him the Aborigines Bennelong and Yemmerawannie, who were introduced to the King of England.

Governor Arthur Phillip (1738–1814) was well respected as a sensible, just man undertaking a difficult task.

the FACTS!

WHEN THE FIRST FLEET sailed into Sydney Cove, Aboriginal people met them on the shore, shouting "Wurra Wurra", which means "Go away!" in the language of the local Aborigines.

THE SHIPS WERE NAMED *HMS Sirius, HMS Supply, Alexander, Lady Penrhyn, Charlotte, Prince of Wales, Scarborough, Friendship, Fishburn, Golden Grove* and the *Borrowdale.*

ALL OF THE SHIPS but the *Sirius* and the *Supply* were hastily unpacked and then set sail back to England, with some taking on a cargo of tea from Asia on the return journey.

MOST OF THE CONVICT women aboard were young, between 16 and 35 years of age.

AT FIRST, convicts lived in tents, but they were quickly put to work constructing more permanent dwellings. Huts were made of wattle-and-daub, which often collapsed in drenching rain.

MORTAR FOR BRICKS was unavailable, so women collected oyster shells to be ground into lime for cement. Despite this, most brickwork failed to set properly and most of the early buildings collapsed!

PHILLIP was missing the tooth that the Aborigines traditionally pulled out in a manhood initiation ceremony – this seemed to help them trust him and relate to him.

Source: Amazing Facts about Australia: *Early Explorers*, pp. 34-35, Steve Parish

Research

Find out five interesting facts about Governor Arthur Phillip.

1 ______________________________

____

2 ______________________________

____

3 ______________________________

____

4 ______________________________

____

5 ______________________________

____

Questioning

Write your own questions that can be answered from the text.

Who ______________________________

____

Why ______________________________

____

When ______________________________

____

What ______________________________

____

Where ______________________________

____

Analysing

What do the images tell you about life in the new colony?

____

____

____

____

____

____

Communicating

What do you think makes a good leader?

Discuss your ideas with others and create a list of five attributes of a good leader.

Evaluating and Reflecting

"Governor Phillip was a worthy leader." Do you agree or disagree with this statement? Explain your answer.

____

____

____

____

Food Rationing

Farming in the early days of the colony was difficult. This was mainly due to lack of knowledge about the hot and dry climate, the reversal of the seasons (due to moving from the Northern Hemisphere to the Southern Hemisphere), and the suitability of the land. Many of the convicts were professional criminals who had not worked as labourers or farmers before. Another problem was that people on the First Fleet had brought shovels, wheelbarrows, fishhooks and even rabbits – but they didn't bring ploughs for farming. Much of the livestock was reserved for breeding stock, so the only available food was the ships' stores (and much of this was rancid due to the long sea voyage).

Colonists had to rely on strict rationing supplemented by whatever native flora or fauna they could hunt or gather. Fish were plentiful and were shared with the Aboriginal people. Fresh fruit and vegetables were in short supply, so native herbs were used to prevent scurvy. The colonists hunted and ate emus and kangaroos, which one convict described as "like mutton, but much leaner", as well as ducks and other birds.

But despite their rationing, nine months after they landed, the colony ran out of seeds, salt pork and flour. In desperate need of supplies, it was decided that *HMS Sirius* would be sent to the Cape of Good Hope in South Africa to fetch provisions. The colony nearly starved while waiting for the ship to return with its supply of flour, salt pork, medicines and seeds of wheat and barley. Yet while these helped, rations were again very low and by November 1789 the colony was facing starvation. Rations were reduced to two-thirds for "every man, from the Governor to the convict". Life in the colony was miserable and spirits were low.

In 1790, the second convict fleet arrived, bringing with it an additional 759 starving and abused convicts, many of whom were very sick. So the supplies that also arrived with the second fleet were not enough to sustain the growing colony. This, combined with a severe drought that began in July 1790, meant that rations continued to be cut back.

Governor Phillip took action to try and make the colony more self-sufficient. In the area around Parramatta, he had land cleared for farming. He also sent out the *HMS Supply* to Batavia to bring back additional food. By the time he returned to England in 1792, the drought had broken and there were more than 800 hectares of new farmland.

Source: Go Facts Australia: *Colonies*, pp. 6–7, Blake Education; *Blake's Australian History Guide*, p. 25, Pascal Press

TARGETING HASS 4 © PASCAL PRESS ISBN: 9781925726053

Research

How was the climate in England different to the climate in the colony of New South Wales?

Questioning

Write definitions for these words.

rationing:

scurvy:

rancid:

mutton:

self-sufficient:

Analysing

The First Fleet didn't bring any ploughs for farming. What impact did this have on the colony?

Communicating

How would you cope with having your favourite food rationed?

Discuss this with a partner and develop some strategies to help you manage.

Evaluating and Reflecting

Can you think of a time when rationing might be needed? What would be rationed and why would its use need to be carefully managed?

UNIT 12

HISTORY

The Macassans

Some of the first people to visit Australia were the Macassans. They were fishers from Sulawesi (part of Indonesia) who sailed to Northern Australia in small dugout canoes called *perahu* or *prau*. They came every year to fish for *trepang* (sea cucumber). Some people think they may have travelled to Australia as early as the 1500s.

The Macassans set up camps on the shores of Northern Australia, where they boiled and then smoked the trepang in large pots to preserve them. The Macassans stayed in Australia for many months until they had enough trepang to take back home to trade with the Chinese.

An illustration from 1845 of Macassans boiling trepang at Victoria, Port Essington

The local Aboriginal peoples, the Yolngu, worked with the Macassans and some even sailed back to Indonesia with them. The Macassans and the Yolngu did not always get along and sometimes fought, but usually things were peaceful between the two groups.

The Macassans traded metal knives, blades and axes as well as smoking pipes and fish hooks in return for tortoiseshell and pearl shell. The metal tools made cutting things such as wood and food much easier for the Yolngu. The Yolngu also liked the Macassans' dugout canoes, so they copied the design to make their own.

Dried and live trepang

The visits of the Macassans are remembered by local Aboriginal peoples in their stories, songs, dances and paintings. Some Macassan words are still used today in Aboriginal languages along the north coast. Some examples include *rupiah* (money), *jama* (work) and *balanda* (white person). The Macassans also planted tamarind trees, which still grow in some parts of Northern Australia.

Source: Australian History Centres: *Middle Primary*, p. 69, Blake Education

TARGETING HASS 4 © PASCAL PRESS ISBN: 9781925726053

Research

The Macassans traded with the local Aboriginal peoples, the Yolngu.

Who are the Yolngu? Where do they live? What customs do they have?

Analysing

Why do you think the relationship between the Macassans and Yolngu continued for such a long time?

Questioning

True or false?

1 The Macassans dugout canoes were called *Sulawesi*.

2 Trepang were preserved by being boiled and smoked in large pots.

3 The Macassans and Yolngu were always friendly.

4 The Yolngu used metal tools.

5 The Macassans gave the Yolngu tortoiseshell and pearl shell.

6 The languages of the Macassans and Yolngu became mixed over time.

Communicating

Discuss the illustration from 1845 of the Macassans at Port Essington. What does it tell you about this group of people?

Share your ideas with a partner or in a small group.

Evaluating and Reflecting

When European settlers arrived on the First Fleet, the Aboriginal Peoples of the Eora nation (the coastal Sydney area) shouted "Wurra wurra", which means 'go away'. The Europeans did not respect the Aboriginal people and they did not understand their culture. There was much conflict.

How does this compare with the interactions between the Yolngu and the Macassans? Why are these differences important?

UNIT 13

Assistance, Resistance and Reprisal

At first, the Aboriginal people were curious about the strange white men and women, and some even helped them out. However, after a while, once they realised these "visitors" were here to take over their traditional lands, naturally Aborigines tried to resist the invasion.

Some Aborigines, such as Bennelong, became distanced from their culture because of their relationship with Europeans.

HISTORY

SOME TOOK to attacking convicts and settlers and spearing livestock, sparking bloody confrontations. Both Aborigines and white settlers were killed in the years of conflict that followed, but many more Aborigines than Europeans lost their lives. Aboriginal culture and beliefs were difficult for the settlers to fathom. They could not understand why the Aborigines did not want to wear clothes or live in houses. The Aborigines, however, could not understand why the white people should need such things. Despite the conflict, some Aborigines and Europeans formed firm friendships and many Aborigines provided invaluable help to the settlers. Often they worked as trackers and guides for explorers; as native police; on farms or as station hands mustering cattle; or working as cooks or domestic servants.

Settlers and Aborigines often came into conflict, although some governors tried hard to provide justice for both parties.

A FISH OUT OF WATER

Bennelong ("great fish") was a tall warrior of the Wanghal tribe who could spear up to twenty fish in an afternoon. In 1789, he and his friend Colbee were caught by Captain Phillip, who wanted to know how to find native food. Colbee escaped, but Bennelong did not seem to mind living in the colony, so his leg-irons were removed. He learned English quickly and called Governor Phillip *Be-anna* - the name given to the most respected elder. In return, Phillip called Bennelong *Dooroow*, which means son. Bennelong later accompanied Phillip to England, where he learnt to skate, box and even met King George III. On his return, he urged his people to adopt the English way of life. Bennelong's sister ran from Botany Bay to greet him, but he was angry that she was not clothed. The Wanghal people were wary of this new Bennelong. He gave one of his wives a bonnet and petticoat as a gift, so she left him. He lost the respect of his tribe. He lived a lonely life in his brick house and later lost the respect of the colonists when he turned to alcohol. Bennelong died an outcast in 1813. Bennelong Point, where the Opera House stands, is named after this friendly Aborigine.

the FACTS!

GOVERNOR PHILLIP believed in the concept of the "noble savage" popularised by philosopher Jean-Jacques Rousseau – that "natives" were naturally more honest and noble than greedy Europeans. He wrote, "I shall think it a great point gained if I can proceed ... without having any great dispute with the natives; a few of them I shall ... settle near us ... to furnish [them] with everything that can tend to civilise them, and give them a high opinion of their new guests."

MOST EUROPEANS couldn't understand the Aboriginal way of life. They hoped to bring Christianity to Aborigines and believed that children who were given a Christian education would be integrated into white society, so they forcibly removed Aboriginal children from their families.

SOME ABORIGINES requested that the white men use their guns against tribal enemies and were upset when they would not. But the Aborigines hated the hangings and floggings they saw meted out by the white men on the convicts.

BRAVE PEMULWUY of the Dharug people gathered a force of 100 warriors and fought a twelve-year battle of resistance with European settlers, making raids at Toongabbie and Parramatta in 1797 and torching farms around the colony.

Source: Amazing Facts about Australia: *Early Settlers*, p. 24, Steve Parish

Research

Pemulwuy of the Dharug people was a resistance fighter. What did he do and what impact did it have on the relationship between the Aboriginal people and the Europeans?

Questioning

Examine the image showing conflict between Aboriginal people and settlers.

Write four questions that you would like answered about the situation.

1

2

3

4

Analysing

Why did Bennelong lose the respect of his tribe?

Communicating

Imagine that a foreign group of people comes into your local area. They look different, wear strange clothes and carry weapons. What would you do?

In a small group, discuss possible ways to respond and examine the positive and negative consequences of these actions.

Evaluating and Reflecting

Do you agree with Governor Phillip that the Aboriginal people were 'noble savages'? Explain your answer.

UNIT 14

Truganini

HISTORY

Read the lyrics of this song featuring Truganini and discuss the significance of music like this: www.azlyrics.com/lyrics/midnightoil/truganini.html.

A LEADER

Truganini was one of the foremost Tasmanian Aboriginal leaders of the 1800s, and a negotiator and spokesperson for her people. She was born in Van Diemen's Land in 1812, and was the daughter of Mangana (Mangerner), the leader of the Recherche Bay people. By the time Truganini was 17, she had experienced the abduction of her sister and the murders of her mother, her fiancé and an uncle.

BLACK WARS

Between 1828 and 1832, many conflicts were fought between Aboriginal peoples and European settlers. These conflicts were referred to as the 'Black Wars'. As a result, the government decided to relocate Aboriginal peoples to island reserves. George Augustus Robinson was responsible for the first of these missions on Bruny Island in 1829, and this was where he met Truganini. Disease eventually caused the closure of the Bruny Island Mission in 1830, but Robinson still believed that he could civilise Aboriginal peoples through Christian teachings.

FLINDERS ISLAND

From 1830 to 1835, Truganini accompanied Robinson on his travels across Tasmania, acting as his guide and interpreter. Robinson and Truganini managed to convince the remaining Aboriginal peoples (numbering about 160 at the time) to relocate to Flinders Island in 1835. Truganini believed that they would be safe and protected from the violence of the conflicts, but many people died from diseases caused by the poor living conditions.

Flinders Island

PORT PHILLIP & OYSTER COVE

In 1839, Robinson became Chief Protector of Aborigines for Port Phillip in Victoria. Truganini and 15 other Aboriginal people accompanied Robinson to Port Phillip. While there, two of the Tasmanians were hanged for murder and the rest were sent back to Flinders Island. Truganini and 46 other survivors were moved to Oyster Cove in 1847. By 1869, Truganini was one of only two full-blood Tasmanian Aboriginal people still alive.

Truganini (far right) with three other Tasmanian Aboriginal people.

THE LAST OF HER PEOPLE

Truganini died in Hobart in 1876, at the age of 64. Before her death, she feared her remains would be used for scientific study or displayed in a museum. Her skeleton was eventually put on display at Hobart Museum until 1947. It was not until 1976, one hundred years after her death, that her remains were cremated and scattered in the waters close to her homeland.

Source: *Blake's Australian History Guide*, p. 37, Pascal Press

TARGETING HASS 4 © PASCAL PRESS ISBN: 9781925726053

Research

Find out about Truganini's life as a bushranger.

Questioning

If you were able to interview Truganini, what would you ask her?

Write at least four questions.

Analysing

Why were these conflicts referred to as the 'Black Wars'? What would have been a more appropriate name? Give reasons for your answer.

Communicating

Using your research, write and present a short speech about Truganini's life.

Include her achievements and the challenges she faced.

Evaluating and Reflecting

Do you think that Truganini is an Aboriginal hero? Explain your answer.

UNIT 15

Weather and Climate Around the World

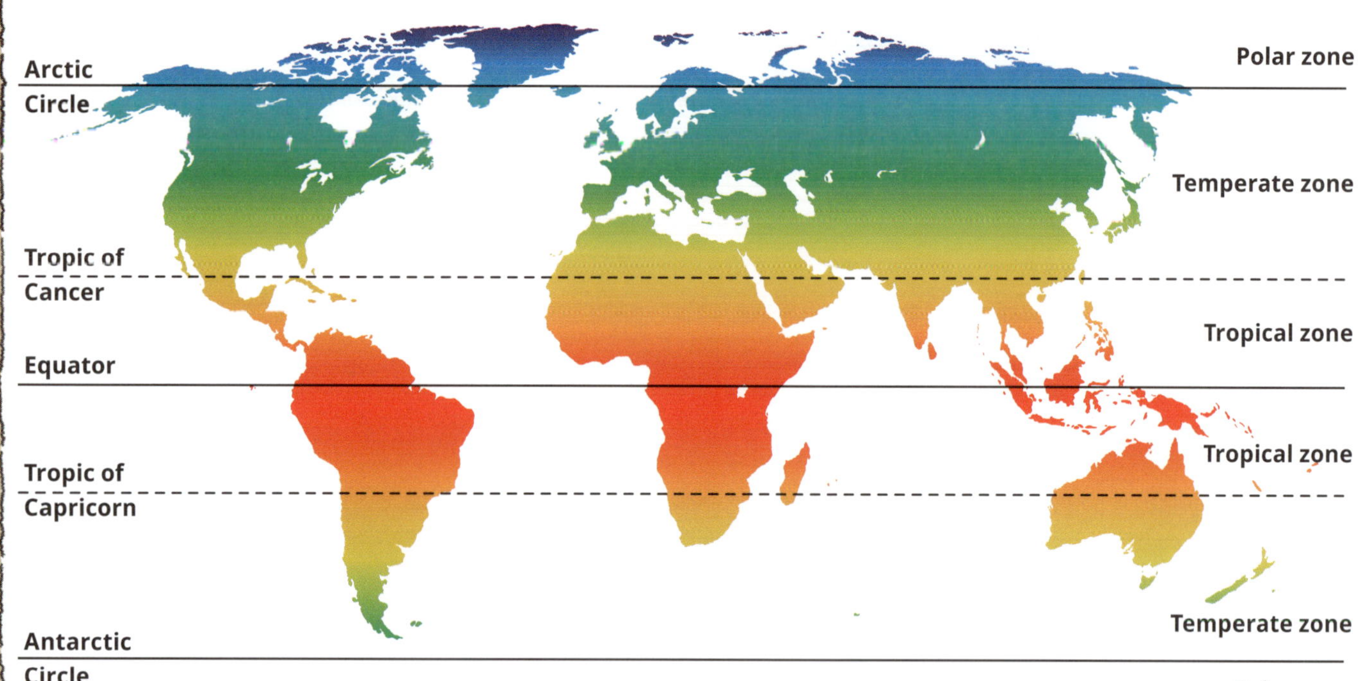

GEOGRAPHY

Tropical zones

The red and orange areas on the map represent the hot or tropical zones of the Earth. The hottest areas are shown in red. These are the places closest to the equator. Some tropical zones have extreme heat and humidity in summer, as well as violent storms and rain. These regular, heavy rains are called monsoons. Winters in the tropics are drier and cooler than the summers, but are still quite warm. Some tropical zones are dry and are composed of desert regions where there is very little rainfall. Deserts can become quite cold at night, even though their daytime temperatures are extremely high.

Temperate zones

The green and yellow areas on the map represent the cool and warm temperate zones. The cooler zones are green and the warmer zones are yellow. In these areas, there are cool winters and warm summers. There may be some snow in winter and some hot days in summer, but temperatures are usually mild across all of the seasons. Temperate zones usually have lots of regular rainfall.

Polar zones

The blue areas on the map represent the polar zones inside the lines of latitude of the Arctic and Antarctic Circles. The climate of these areas is cold, snowy and icy with little rainfall.

Source: Australian Geography Centres: *Middle Primary*, p. 17, Blake Education

TARGETING HASS 4 © PASCAL PRESS ISBN: 9781925726053

Research

Find out where Antarctica is and what it looks like. Draw it on the map and label it.

What is the torrid zone? At what latitude is it found?

____

____

____

List at least three interesting things about the torrid zone.

____

____

____

____

____

____

Questioning

1 Which major line of latitude crosses Australia?

____

2 How are polar zones similar to deserts?

____

____

3 Which zone is most of Africa in?

____

4 Where would you find the weather to be very hot during the day and cold at night?

____

5 What sort of weather conditions would you expect to have in New Zealand, and why?

____

____

____

Analysing

What happens to the climate as you move away from the equator? Why does this happen?

____

____

____

____

Communicating

Imagine that you are video calling a friend from your holiday destination in another country.

Describe the climate and weather you are experiencing. Can your friend guess where in the world you might be?

Evaluating and Reflecting

What type of climate do you think is the best to live in? Why?
Describe how this climate supports the survival of plants and animals.

____

____

____

____

UNIT 16

South America

Like Australia, South America is a large continent. It is made up of 12 countries and has a population of 422.5 million people. The total land size of South America is 17.84 million square kilometres. It is connected to the continent of North America by the Isthmus of Panama.

In the north, the Amazon River Basin is the largest area of tropical rainforest in the world. Its forests are filled with liana vines. Lianas use tendrils or suckers to grasp young tree saplings and the branches of mature trees. This allows them to climb towards the forest canopy. Once there, lianas grow between trees and form a strong web which helps to stabilise trees in strong winds. Liana webs are vital to the survival of animal species in the ecosystem. They allow other animals to move through the forest canopy, searching for food, without needing to visit the forest floor, where their predators live.

Did you know?

Almost 400 billion trees belonging to 16 000 different species grow in the Amazon Basin.

Lianas form tight masses of growth.

The rivers of the Amazon Basin hold more fish species than the Atlantic Ocean. Perhaps the most famous South American fish is the piranha. There are more than 30 species of piranha. Many of them are omnivorous, feeding on seeds and meat. They are scavengers with an excellent sense of smell. They eat weak or dead fish and animals which would otherwise die and pollute their ecosystem.

In the centre of South America, the Pantanal is the world's largest freshwater wetland. In the west is the Atacama Desert, one of the driest places on Earth. Many plants there have long tap roots that retrieve water from deep underground.

Further south, the Patagonia region is home to armadillos and burrowing parrots. It is also the last wild environment of the guanaco, a member of the camel family. At the southern tip of the continent, Tierra del Fuego features beech forests, peat bogs, alpine meadows, rivers and fjords.

People damage South America's ecosystems through road building, overhunting and clearing land for farming. Farming in the Pantanal pollutes wetlands and changes the flood cycles. In Tierra del Fuego, beavers - introduced from Canada in 1946 - cause great damage to forests and rivers.

Sloths rarely go to the rainforest floor.

Source: Australian Geography Centres: *Middle Primary*, p. 31, Blake Education; Go Facts Geography: *Ecosystems*, pp. 12-15, Blake Education

GEOGRAPHY

TARGETING HASS 4 © PASCAL PRESS ISBN: 9781925726053

Research

What is an isthmus? Where around the world are isthmuses found?

Questioning

Write four questions you have about South America after reading the text. Can you find the answers to these questions?

1

2

3

4

Analysing

NASA has used the Atacama Desert of South America for testing conditions for future missions to Mars.

What would be some of the advantages of choosing this place for the testing?

Communicating

Take on the role of an adult sloth.

You need to explain to a young sloth why they have to stay in the liana vines.

What would you say to them?

Evaluating and Reflecting

Why is it important to prevent damage to South America's ecosystems? How can people do this?

UNIT 17

Africa

GEOGRAPHY

Divided by the equator, Africa features many diverse ecosystems and is the second largest continent in the world. It is made up of 54 countries and has a population of 1.2 billion people. The total land size of Africa is 30.37 million square kilometres. It is connected to the continent of Asia by the Isthmus of Suez in Egypt.

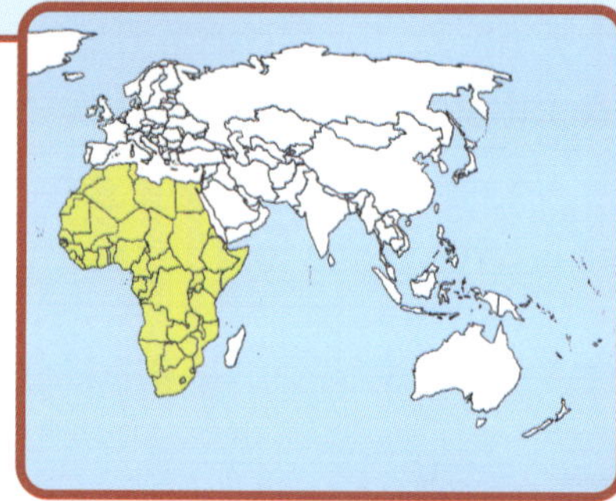

Dominating northern Africa is the Sahara, the largest hot desert in the world. Central Africa features the Congo River Basin. Forests around the Congo River are home to some 400 mammal species, 1000 bird species and more than 10 000 plant species – one third of which are found nowhere else.

The whistling thorn is a tree native to African woodlands and rainforests. It has long thorns on its branches, each with a hollow bulb. The thorns help to protect the plant from grazing animals, such as giraffes and elephants. The whistling thorn has a symbiotic relationship with stinging ants. Its hollow bulbs are homes for the ants, and it provides nectar for them to eat. In return, the ants help protect the tree by swarming out and stinging grazing animals – even elephants.

About half of Africa is savanna. Grasslands dotted with trees are home to cheetahs, lions, elephants, gazelles, and many other species. They are the location of the annual migrations of wildebeest and other large animals.

Huge lakes in the east of the continent are some of the oldest in the world. Lake Malawi has the richest freshwater fauna in the world – about 14 per cent of the world's freshwater fish species.

Southern Africa has a temperate climate but also has deserts and wetlands. The Okavango Delta has permanent wetlands and plains that flood each year. The Namib Desert is a sand desert along the South Atlantic coast.

Ants put holes in the hollow bulbs. The whistling thorn gets its name from the sound of the wind blowing over old bulbs.

Source: Australian Geography Centres: *Middle Primary*, p. 31, Blake Education; Go Facts Geography: *Ecosystems*, pp. 12-15, Blake Education

TARGETING HASS 4 © PASCAL PRESS ISBN: 9781925726053

How does the environment support the lives of people and other living things?

Research

Select one African country.

Write a brief information report about its climate and significant natural features.

Questioning

Use the text to help you write questions that have these answers.

Answer: savanna
Question:

Answer: Okavango Delta
Question:

Answer: Suez
Question:

Answer: Sahara
Question:

Analysing

A symbiotic relationship is a close relationship between two living things that benefits one or both of them. How is the relationship between stinging ants and whistling thorns symbiotic?

Communicating

Would you like to visit the African country you researched earlier? Explain your reasons and give evidence.

Evaluating and Reflecting

Many of Africa's iconic animals such as lions, elephants and rhinoceroses are on the edge of becoming extinct due to human actions. What do you think will happen to the fragile ecosystems of Africa if these animals do become extinct? How will other plants and animals be affected?

Australia's Living Resources

Noongar country is located in south-west Western Australia and includes the city of Perth. This region of Australia is known for its different landscapes. These include woodlands, swamplands, coastal plains and forests. These different environments mean that there is a lot of variety in the types of vegetation found throughout the seasons.

Noongar name: Biara
Common name: Candle Banksia

The spikes are boiled in water to release the nectar and make a sweet drink. The dried cone of the flower is used as a firestick to transport fire between camps.

Noongar name: Bain
Common name: Coastal Pigface

The sweet fruit of this plant is eaten fresh or dried. The leaves can also be eaten. The juice from the leaves is used for burns, scalds and stings.

Noongar name: Balga
Common name: Grass Tree

The leaves are used to thatch roofs, and the resin is mixed with charcoal and kangaroo droppings to make glue for tools and weapons. The flowers are soaked to make a sweet drink, and the spike is used as a firestick.

Noongar name: Marri
Common name: Red Gum

The red gum that comes from this tree is used as an antiseptic. It is also used to ease the pain of toothaches, upset stomachs, and throat and gum infections.

Noongar name: Dumbari
Common name: Quandong

This delicious fruit is also known as a wild peach. Its nuts are used to make jewellery or for children's games.

Noongar name: Kara
Common name: Milkmaids

The tubers of this plant taste similar to potatoes. They are baked, roasted or pounded into a paste.

Source: Australian Geography Centres: *Middle Primary*, p. 43, Blake Education

GEOGRAPHY

TARGETING HASS 4 © PASCAL PRESS ISBN: 9781925726053

Research

What are the similarities and differences between woodlands, swamplands, coastal plains and forests? Briefly describe each landscape.

Landscapes	Landscape description
Woodland	
Swampland	
Coastal plain	
Forest	

Questioning

True or false?

1 Red gum sap can be used as an antiseptic.

____

2 Milkmaids can be roasted like potatoes.

____

3 It is harmful to eat dumbari. __________

4 Only the leaves of the coastal pig face can be eaten. __________

5 Balga is a very versatile plant. __________

6 If you have sunburn, then the quandong is helpful. __________

7 Fresh biara is used to carry fire from camp site to camp site. __________

8 Similar environments mean that there is a lot of variety in the vegetation.

____

Analysing

Why is it important to the Noongar people to use many parts of each plant?

____

____

____

____

Communicating

A haiku is a Japanese 3-line poem written about nature. The first and last lines have 5 syllables. The middle line has 7 syllables. The lines often do not rhyme.

Select one of Australia's living resources from the text and write a haiku poem about it.

Evaluating and Reflecting

If there are so many edible types of vegetation in Australia, why do you think early European explorers such as Burke and Wills died of starvation?

____

____

____

____

The Land Affects Us

GEOGRAPHY

Environments influence where and how people live.

> *Humans don't just leave our footprints on the Earth - the Earth leaves a large one on us.*
> Geographer Dr William R. Burnside

Did you know?
Eighty-five per cent of Australians live within 50 kilometres of the coast.

People live in environments that most readily meet their basic needs. Although human factors play a part, such as the location of work, where people live is mainly based on physical factors:

- **climate** - people prefer temperate climates, and extreme temperatures make it difficult to grow food
- **natural resources** - fresh water is needed, plus energy sources (e.g. coal) for industry
- **relief** (high areas) - mountainous areas are difficult places to build in and grow food
- **soils** - fertile soils in river valleys are needed to grow food.

Rivers play an important role in the economics of communities. The Mekong River flows through six countries in Asia. Along its lower stretches alone, about 20 million people live no further than five kilometres from the river. It provides them with transport and food (fish and water for growing crops).

Fish from the Mekong is the main source of protein in people's diets.

A coastal people

Australia's capital cities are all located on the coast and close to major rivers. They have reasonably stable water supplies, which is important in a dry continent such as Australia. Captain Arthur Phillip moved Australia's first British settlement away from Botany Bay because it was a poor harbour and lacked fresh water.

The Murray-Darling Basin is the catchment of the Murray, Murrumbidgee and Darling Rivers. It contains very little water for its size, yet like the Mekong River it is vital for food production. The Murray-Darling Basin contains 40 per cent of Australia's farms and produces one-third of its food, plus food for export.

Melons grown in the Murray-Darling Basin. The average yearly flow in the Basin is less than the daily flow of the Amazon River.

World Population Density

Persons per km²
0
1 - 4
5 - 24
25 - 249
250 - 999
1,000 +

Robinson Projection
Based on 2.5 arc-minute resolution data

Note the low numbers of people living in arid areas (e.g. much of Australia and the Sahara in Africa) and at high latitudes (near the North and South Poles).

Source: Go Facts Geography: *Re-shaping Environments*, pp. 28-29, Blake Education

Research

Find out more about the Murray-Darling Basin and how this important river system is used by different groups in the community.

You can scan the QR code to take you to the Murray-Darling Basin Authority website.

___

___

___

___

___

___

___

___

___

Questioning

Write four *Did you know?* facts based on your research into the Murray-Darling Basin.

1 ___

___

___

2 ___

___

___

3 ___

___

___

4 ___

___

___

Analysing

Look at the World Population Density map.
Why do you think so few people live in arid regions?

___

___

___

___

Communicating

Create a collage of images of the Murray-Darling Basin to illustrate its diverse landscapes and ecosystems.

Evaluating and Reflecting

Why do you think it is important for Australia to protect river systems like the Murray-Darling Basin?

___

___

___

___

Guarding Ecosystems

GEOGRAPHY

People must help to maintain healthy ecosystems.

Almost half of the wild orang-utans live in forests managed by timber, palm oil and mining companies.

Did you know?
The orang-utan is the world's largest tree-climbing mammal.

Ecosystems are delicate. Living and non-living things rely on each other. If one part of an ecosystem changes, it can put the whole ecosystem out of balance. For example, in Canada's alpine forests, people hunted wolves to protect their farm animals. With fewer wolves, the numbers of elk increased. Elk eat the new shoots of willow trees, so tree numbers fell. This caused the numbers of birds to fall, as they had fewer trees in which to nest.

Damaging ecosystems can push plant and animal species towards extinction. In the tropical forests of Indonesia and Malaysia, people log trees which have taken hundreds of years to grow. They clear forests to grow palm trees to extract palm oil. The forests are the habitats for many species, including the orang-utan. Numbers of these apes have fallen as their habitat disappears. Orang-utans spread the seeds of important forest vines. If the orang-utans disappear, it will affect many other plants and animals.

Protecting ecosystems protects life on Earth – including human life. It gives us, and the people to come after us, a rich diversity of plants and animals to enjoy and benefit from.

Forest cleared to grow palm oil trees in Borneo, Malaysia

Source: Go Facts Geography: *Ecosystems*, pp. 20-21, Blake Education

TARGETING HASS 4 © PASCAL PRESS ISBN: 9781925726053

Research

Find out about the social and environmental impacts of palm oil production. Write at least four important facts about this issue.

Questioning

1 What is an ecosystem?

2 Why have numbers of orang-utans fallen in Malaysia?

3 What effect does forest clearing have on ecosystem diversity?

4 What important role do orang-utans play in protecting the diversity of their ecosystem?

5 Where are palm oil plantations found?

Analysing

What is the purpose of the image showing cleared forest in Borneo? How does it impact the reader?

Communicating

Create a flow chart to show the flow-on effects of people hunting wolves in Canada's alpine forests. Share and discuss it with others.

Evaluating and Reflecting

What actions can you take to protect the ecosystems where you live?

UNIT 21

GEOGRAPHY

Extreme and Unique Environments

Some of Australia's wild habitats are extremely specialised or exist in such small proportions that the fauna residing in them have little hope of survival if environmental pressures encroach upon their homes. Alpine and hypersaline habitats are two such environments.

Adelie Penguins in Antarctica are now being affected by decades-old DDT that is seeping out of melting ice.

FROZEN IN TIME

The Australian mainland has very small sections of snow-covered habitat because snow falls on less than 2% of the land surface. Areas where snow remains for more than 30 days are even scarcer, yet some plants and animals rely on this snow cover. CSIRO studies suggest that snow cover in the Snowy Mountains could shrink as much as 95% by 2070 if global warming is not controlled. Australia's offshore frigid areas, such as Macquarie Island and Antarctic Territory, are set to fare little better if catastrophic climate change occurs.

A SLEEPY SNOW-LOVER

The tiny, endangered Mountain Pygmy-possum (*Burramys parvus*, right) is confined to alpine areas where it hibernates under a dense cover of snow and vegetation for up to seven months a year. Three distinct populations occur – at Mt Buller, Mt Hotham and Mt Kosciuszko – and are in decline. In total, its range is less than 10 km^2, much of which is seasonally inundated with skiers or altered by snow-making. In summer, Mountain Pygmy-possums are heavily reliant on a diet of berries, along with feasting on the migratory Bogong Moth, to store up the fat reserves they require for hibernation. In winter, they need snow cover of at least 1 m to cover their nests. At Healesville Sanctuary, Victoria, captive possums from the Mt Buller population, which has dwindled to just 30 wild individuals, are being kept in a refrigerated facility that simulates the possum's natural habitat. There are less than 2000 of these tiny mammals in existence. Snowline and feldmark plants may be at risk of extinction if temperatures put an end to winter snow in Australia's High Country.

SUPER SALTY HABITATS

All soils contain some salt. It is produced by weathering of rock, or is a remnant of earlier geological periods when salt water covered the land, or is 'atmospheric' salt that has been deposited by rainfall (known as cyclic salt). Saline soils occur when there is such a quantity of salt that plants find it difficult to grow. Many species struggle to survive around salt lakes and salt marshes, but others, such as the Lake Eyre Dragon (above) have capitalised on the 'niche' habitat.

the FACTS!

IN 1959 the Antarctic Treaty preserved the frozen continent's pristine wilderness as a place of global importance where scientific research could flourish. In 1998, an environmental protection protocol came into force to protect Antarctica from the effects of mining or drilling for oil for at least 50 years. However, even now, some parts of Antarctica remain at risk from pollution around scientific research stations, airstrips being built too close to penguin colonies, and the impact of earlier drilling.

A COLONY of Little Penguins (*Eudyptula minor*) at North Sydney Harbour – the only such colony on mainland NSW and therefore listed as endangered in that State – is being protected by a recovery program at Taronga Zoo. The program aims to increase the population by enhancing community awareness, protecting habitat and researching these birds.

Source: Amazing Facts about Australia: *Wildlife Conservation*, p. 46, Steve Parish

TARGETING HASS 4 © PASCAL PRESS ISBN: 9781925726053

How do different views about the environment influence approaches to sustainability?

UNIT 21

Research

Collect information on the habitat and adaptations of the Lake Eyre Dragon.

Analysing

Why do extremely specialised environments increase the risk for animals and plants that live there?

Questioning

1 What are the environmental pressures faced by extreme habitats?

2 How is salt produced in soils?

3 Where does the Mountain Pygmy-possum live?

4 What program is running in North Sydney Harbour?

5 What is a 'niche' habitat?

Communicating

In a small group, discuss how mining companies and scientists might view the natural environments and resources of Antarctica in different ways.

Evaluating and Reflecting

Should mining and drilling be allowed in Antarctica once the 50-year environmental protection protocol ends? Explain your reasons.

Ancient Land Management Practices

Australia's Aboriginal and Torres Strait Islander peoples (Indigenous Australians) enjoy a special relationship with the land and the natural resources of the continent. Before the arrival of Europeans in the 1700s, there were many groups of Indigenous peoples living in many different parts of Australia. These groups practised various sustainable land management methods, some of which are still in use today.

Seasonal migration

Indigenous Australians obtained their resources through fishing, hunting and gathering. Traps and nets were used to catch living creatures for food, including fish. By moving from one area to another and using only what was plentiful according to the seasons, Indigenous Australians ensured that the resources in one particular area were never fully used up.

Painting of Aboriginal peoples fishing, ca. 1817

Vegetation

It is believed that some Indigenous Australian peoples scattered seeds of certain plants to ensure that they would be widely available in multiple areas. Indigenous Australians also practised a method known as fire-stick farming. Controlled fires were used to burn away the dry leaves and bushes, and to encourage new plant growth.

Water

Water was collected and saved during times of heavy rainfall. It could be found in tree hollows and tree roots underground. Grazing animals were also tracked and followed to new waterholes.

Animals

Following the migration patterns of animals like birds and fish allowed Indigenous Australians to know what food sources would be available at certain times in certain areas.
Totems are given to Aboriginal and Torres Strait Islander peoples when they are born. A totem is an animal, plant or object in nature that has a special meaning and connection to a family or group. Indigenous Australians usually do not kill or eat the animal of their totem. This ensures that some animals are protected from hunting in certain areas and can continue to thrive there.

Source: Australian Geography Centres: *Middle Primary*, p. 49, Blake Education

TARGETING HASS 4 © PASCAL PRESS ISBN: 9781925726053

Research

Find out about three species of migratory Australian birds or fish. What are their migration patterns?

Species 1: ______________________________

____

____

____

Species 2: ______________________________

____

____

____

Species 3: ______________________________

____

____

____

Questioning

True or false?

1 Aboriginal people always hunted and fished in the same area.

____

2 Fire-stick farming was used to encourage new plants to grow.

____

3 Aboriginal people often eat the animal of their totem. ______________

4 Aboriginal people fished with nets and spears. ______________

5 Aboriginal and Torres Strait Islanders would use all the resources in one area and then move somewhere else.

____

6 Indigenous Australians used water that had collected in tree hollows.

____

Analysing

In what ways are ancient Aboriginal land management practices sustainable?

____

____

____

____

Communicating

Create a map of Australia and its near neighbours to show the migration patterns of the birds or fish you researched above. Use different colours to show different species.

Evaluating and Reflecting

Are the ancient land management practices still useful in Australia today? Explain your reasons.

____

____

____

____

UNIT 23

Connecting to Land

Land means different things to different people. For Aboriginal peoples, land is the home of their spirit ancestors – they are spiritually, socially and culturally tied to it. It gives them their identity.

Relationships with the land influence how people think it should be used. To a property developer, land is something to buy and sell to create wealth. For a farmer, the land is both a home and a resource they use to make a living. An environmentalist believes healthy land is vital to human survival and so they try to protect land from damage by human activities.

Did you know?
In a Torres Strait Islander creation story, a mother and son, Nageg and Geigi, became the triggerfish and the great trevally.

Land brings peace and joy.

For Aboriginal and Torres Strait Islander peoples, the land is not just dirt, rocks and water. Aboriginal peoples feel a strong sense of belonging to the land – their country. They believe that when their ancestral spirits first moved across the land, they created animals, plants and people. They also created landforms, such as mountains and rivers. This is the Dreaming. When the spirits finished, they changed into waterholes, mountains and other places. Such places are sacred to Aboriginal people.

For the Gagudju people of the Northern Territory, life began when Warramurrungundji, the Fertility Mother, moved over the land, planting food and digging waterholes. When her journey ended, she sat down and changed into a large rock. This marks her Dreaming site.

Land brings food and life.

People talk about country the same way they would talk about a person: they speak to country, sing to country, visit country, worry about country, feel sorry for country and long for country.

Social scientist Deborah Bird Rose

Lucy Yukenbarri paints her country in the Great Sandy Desert – waterholes, dunes and bush foods.

Art is one way Aboriginal people express their connection to country. By painting their country, they are passing on their stories. Rover Thomas (1926–1998) painted his desert country and the Kimberly area where he lived. Thomas's paintings are like maps. They show riverbeds, tracks, valleys and hills. They are also spiritual maps that tell Dreaming stories. Painting reconnected Thomas with his ancestors and the land.

Source: Go Facts Geography: *People and the Land*, pp. 4-8, Blake Education

GEOGRAPHY

TARGETING HASS 4 © PASCAL PRESS ISBN: 9781925726053

Research

Examine one Aboriginal Dreamtime creation story, such as *Tiddalick the frog, The rainbow serpent* or *How the birds got their colours.*

What does the story tell you about the Australian environment?

Questioning

If you could talk to Lucy Yukenbarri about her art, what would you ask her?

Write at least four questions.

Analysing

How does the Aboriginal peoples' spiritual connection to the land influence their approach to using it sustainably?

Communicating

Create your own artistic map of your school or local area. Include special natural and human features.

Evaluating and Reflecting

Do you have a special connection with a particular place? It might be a local park, a favourite holiday destination or a quiet spot, just for you. How does your connection influence your use of the place?

UNIT 24

Reshaping Environments

People have the ability to reshape - both damage and improve - natural environments.

A natural environment is the land, air, water and climate in which organisms live. People live in many types of environments - and they change them. They build dams that change the course of rivers. They dredge harbours, which changes the water conditions for marine life. They clear land for farming and to build houses, which removes the habitats of living things.

People can sometimes repair environments that they damage. They can remove dumped rubbish and plant new trees. However, some activities, such as open-cut mining, change the environment so much that it can never be restored to its original state.

Replanting logged old-growth forests with new trees does not guarantee animal species will return to their environments.

Tree roots bind the soil. If you remove trees to clear land, landslides can occur.

People pollute environments - and only sometimes remove the pollution.

Did you know?

The Chinese river dolphin - found only in the Yangtze River - became critically endangered and possibly extinct after the Three Gorges Dam was built.

Three Gorges Dam

The Three Gorges Dam across the Yangtze River in China generates electricity for millions of people. By controlling the river's flow, it allows greater trade along the river. It also helps control the Yangtze's massive floods, which have killed thousands of people and caused billions of dollars of damage.

However, the dam has reshaped environments. The project flooded hundreds of factories, mines and waste dumps, adding industrial pollution to the environment. It forced more than one million people to find new homes and jobs. Downstream from the dam, erosion caused landslides. Many fish and plant species were not suited to the new flow, temperature or depth of the river. Hundreds of kilometres from the dam, changes to the river's depth and water quality have affected the food supply of rare crane species.

Source: Go Facts Geography: *Reshaping Environments*, pp. 4-5, Blake Education

GEOGRAPHY

TARGETING HASS 4 © PASCAL PRESS ISBN: 9781925726053

Research

Find out more about the Chinese river dolphin.

Write at least five interesting facts about it and its changing environment.

Questioning

Write at least four questions that a friend can answer by reading the text.

Analysing

What did you find most interesting when reading the text? Why?

Communicating

In a small group, discuss ways to minimise the negative impacts of erosion when clearing land for new developments.

Evaluating and Reflecting

What conclusions can you draw about the environmental impact of the Three Gorges Dam? Explain your response.

Energy from the Earth

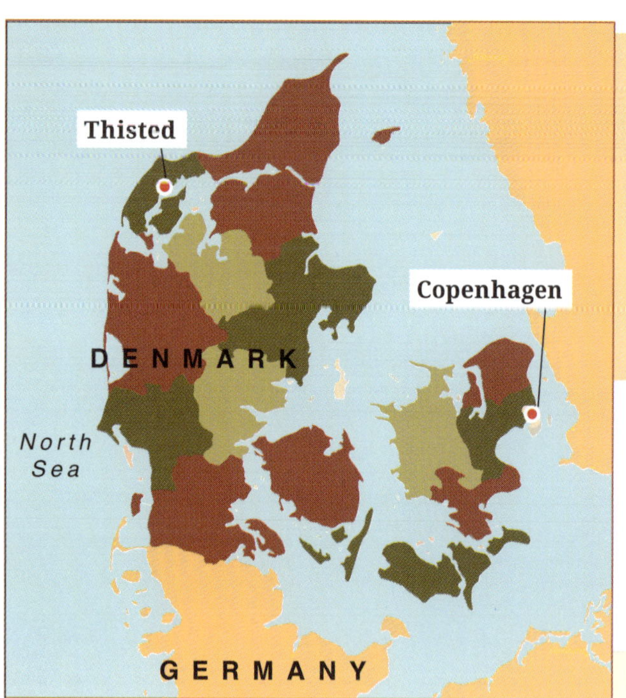

A municipality in Denmark is producing clean energy from renewable resources. The people of Thisted have found a great solution to their energy problems. This community of around 46 000 people is an inspiring example of how working together and using common sense can bring clean and green power to the people.

Thisted, Denmark

Wind turbines

Thisted produces 100% of its electricity and more than 85% of its heating through clean and renewable means. Because of Thisted's position on the windy North Sea, wind power is used instead of coal and oil-burning power stations. But wind power is not the only resource that has enabled Thisted to switch from fossil fuels to clean energy.

They use two common waste products that everyone wants to get rid of — straw and household garbage. These products are collected from farms and residential areas, and are used to fire power plants. This leads to low-cost energy for everyone and less waste going to landfill.

The people of Thisted are always looking for new ways to produce clean energy. They have programs for solar power, wind energy, recycling, agricultural and forestry waste, wave and tidal energy, and geothermal (underground) heat. This community acts as a wonderful example to the rest of the world that using totally clean energy is an achievable goal.

Source: Australian Geography Centres: *Lower Primary*, p. 31, Blake Education

TARGETING HASS 4 © PASCAL PRESS ISBN: 9781925726053

Research

Choose one form of clean energy – solar, wind, geothermal or wave/tidal.

Explain how this form of energy is produced.

Questioning

What questions do you have about renewable energy? Write at least four questions.

Analysing

Why is what Thisted does 'a wonderful example to the rest of the world'?

Communicating

Create a visual flow chart showing the steps in generating the form of clean energy that you researched earlier. Remember to label your diagrams.

Evaluating and Reflecting

Should your community create energy in the same way that the community of Thisted does? What would be the benefits and drawbacks?

Living Sustainably to Conserve Wildlife

Sustainable development aims to help people to find ways to live and work that do not threaten the environment. By doing this, we can make sure that wildlife and humans both benefit from conservation and can coexist harmoniously.

Aquaculture, such as this salmon farm in Tasmania, may help relieve some of the pressure on wild populations of fish and crustaceans.

CARRYING THE WEIGHT OF THE WORLD

Sustainable development promotes the uses of wildlife and habitats for human benefit, but in a way that makes sure they are maintained for future generations. The Earth's sustainability depends on its carrying capacity – how many organisms and humans each area can support without degradation of that region. Some researchers have calculated that the Earth's capacity is just 3 billion people – by this measure, humans are already at more than twice their sustainable level! Others suggest that the planet is capable of carrying as many as 44 billion people. The figure is difficult to work out because different peoples place different demands on the environment. Just 20% of the world's population consumes 80% of the world's resources.

TREADING LIGHTLY

An 'ecological footprint' is calculated by measuring how much food, energy and other materials humans use, and then calculating the land and sea required to produce these needs and to absorb any waste that results from their production and use. When WWF (World Wildlife Fund) measured the globe's ecological footprint in 1997, they found it was at least 30% larger than the planet's productive areas! By this measure, the planet simply cannot produce resources fast enough to support the human population. An ecological footprint is measured by how many global hectares (each about 2.5 acres) are needed for sustainable living. For each person, 1.9 global hectares is considered sustainable, but many countries exceed this and have a much larger footprint. In the UK, the ecological footprint is 5.35 global hectares per person, and in the USA it is 9.70. Australians have one of the largest ecological footprints at around 2.5 times that of a person in the USA. This is in part because resources are transported for each Australian citizen annually. A study produced in 2007 indicates the world is failing in its efforts to be sustainable. Generally, the more developed a country, the bigger its ecological 'boots' become.

the FACTS!

OVER 100 WORLD LEADERS met in Johannesburg, South Africa, in 2002 to discuss sustainable development. Aims were to provide 10% energy from renewable resources by 2010, to close the gap between rich and poor countries, to use more sustainable products when creating goods, and to restore and protect global ecosystems. Unfortunately, there was little agreement on how to make this happen.

THE MOST EFFECTIVE way to encourage people to practise sustainable development is to introduce them to nature's treasures. Conservation groups are aware that people are more likely to protect an environment when they can see its value. Tourism is one way of protecting the environment while earning a living from it, but it must be sustainable. If too many tourists flood to a wilderness area, it may force the animals that are attracting the tourists to change their lifestyles.

Source: Amazing Facts about Australia: *Wildlife Conservation*, p. 23, Steve Parish

GEOGRAPHY

TARGETING HASS 4 © PASCAL PRESS ISBN: 9781925726053

Research

What is ecotourism? Why is it important? What are its benefits?

Questioning

Fact or opinion?

1 An ecological footprint is calculated by measuring how much food and materials humans use.

2 Humans are constantly destroying the environment.

3 Tourism can never be sustainable.

4 World leaders set unrealistic goals at their 2002 meeting.

5 Aquaculture is practised in Tasmania.

6 Australia has a large ecological footprint.

Analysing

Why does Australia have one of the biggest ecological footprints in the world?

Communicating

Create an outline of your footprint. Inside it draw or paste images of the Earth's resources that you use on a regular basis.

Evaluating and Reflecting

What can people do to reduce their ecological footprint? Should they take such action? Explain your answer.

Understanding your Local Council

In Australia there are three levels of government:

- Federal
- State
- Local.

Local governments are usually called city councils or shire councils, representing their geographical area. In 2019 there were 537 local councils across Australia. However, there are no separate local governments in the Australian Capital Territory. Every three or four years, elections are held to elect representatives to the local council. These people are called councillors, while the leader of the council is known as the mayor.

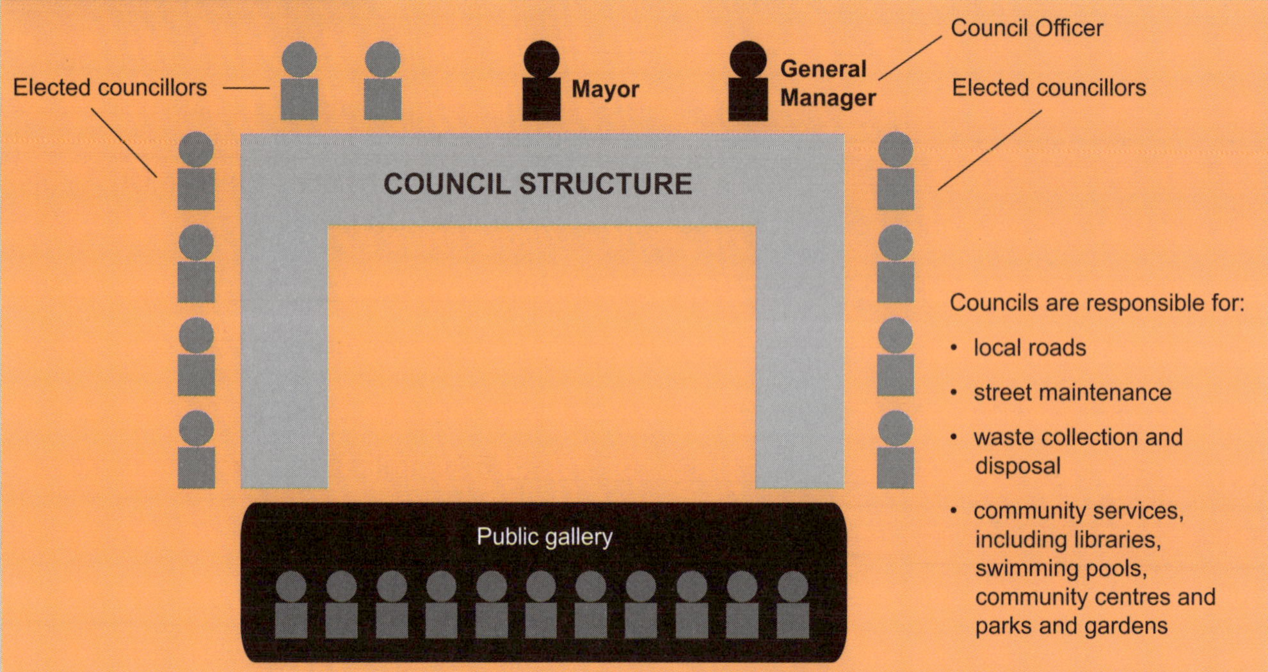

The roles and responsibilities of local councils differ from state to state, but include:

- maintaining local roads, bridges and footpaths
- waste collection and garbage disposal
- providing sport and recreational facilities such as playgrounds, golf courses, swimming pools and sporting ovals
- community services such as childcare/ preschools, aged care homes and some welfare services
- operating libraries, marinas and cultural centres
- water and sewerage services
- monitoring parking facilities
- approving some buildings and new developments.

The decisions made by the local council are called by-laws and only apply to the council that made the law. Councils make by-laws to respond to important local issues and community needs. They often protect public health and safety and try to make sure that the actions of an individual or group do not adversely impact others in the community.

Council revenue comes from collecting rates, a type of tax, which are paid by people who own land in the area. State and federal governments also provide local councils with funds through financial assistance grants for special projects.

Source: *Living and Work Literacy 2*, p. 8, Blake Education

CIVICS AND CITIZENSHIP

TARGETING HASS 4 © PASCAL PRESS ISBN: 9781925726053

Research

Find out about your own local council.

What is your local council area called?

Who is the mayor?

What by-laws have been passed recently?

What local facilities is the council responsible for?

Where and when does the council meet?

Questioning

If you attended a meeting of your local council, what four questions would you ask about the council's contribution to community life?

Analysing

What do you think are the five most important responsibilities of local governments? Rank them in order, starting from the most important. Explain your choices.

Communicating

Discuss your top five local council responsibilities with a friend. Negotiate with them to create a combined top five.

Then get together with another pair and repeat the process. What is your final list? How difficult was it to reach an agreement?

Evaluating and Reflecting

Using your own local council research, are you happy with what the council is doing to help the local community? Explain your answer.

Local Council Case Study

Wagga Wagga is a major regional city located in south-eastern Australia on the Murrumbidgee River. Its local government area population is around 66 000 residents across an area of 4825 square kilometres.

In 2019, the Mayor of Wagga Wagga council was Greg Conkey. His responsibilities included representing the council at public events, welcoming important dignitaries to the city and being the chairperson of council committees, as well as making policy decisions when the council was not meeting.

The Wagga Wagga city council spent more than $75 million on major projects in 2019–2020, including roads, upgrading the sewerage system and park maintenance. The projects are organised through the council's Delivery Program and Operation Plan.

One of the council's projects is their Active Travel Plan. This involves creating 51 km of dedicated cycle paths to go all over the city, providing residents with a safe way to travel to school and work. This will reduce the need for people to use their private vehicles for travelling short distances. These cycle paths will also link the various sporting facilities across Wagga Wagga, with the aim of enhancing the liveability of the city and encouraging people to visit.

Before they began the Active Travel Plan project, the city council discussed it with members of the community. During the design phase, Wagga Wagga residents get to have their say about proposed routes, safety issues and land ownership. Because it is a large project, the Active Travel Plan is implemented in collaboration with the NSW government's transport department.

TARGETING HASS 4 © PASCAL PRESS ISBN: 9781925726053

Research

Research one project being implemented by your local council.

What is it? Who will use it? How does it contribute to community life?

Questioning

Who will benefit from the Active Travel Plan in Wagga Wagga?

How will they benefit from the plan?

Which levels of government are involved in the plan?

What types of tourists is Wagga Wagga hoping to attract with their plan?

Who might be disadvantaged by the plan? Why might they be disadvantaged?

Analysing

Do you think the Active Travel Plan is a good idea? Why?

Communicating

Draw a map of your local area. Indicate where the best place for a bike path would be. Share your ideas in a small group.

Evaluating and Reflecting

Would a similar Active Travel Plan work in your local area? Would your council be able to encourage people to ride bikes instead of driving cars? Explain your response.

Laws and Rules

Laws

Laws are sets of rules created by local, state or federal governments to make sure that individuals and groups within a community behave in an appropriate way. Laws apply to everyone in the community, even if you don't live there and are only visiting or on holidays. They are there to protect people, places and property.

Some laws protect us. For example, the law says we must wear a seatbelt when in a car. So if there is an accident, this law will help to keep us safe.

Some laws protect others. For example, the law says that dogs must be on a leash when walking in public places. So if a dog becomes aggressive, it is less likely to run off and harm someone.

Some laws protect places. For example, there is a law that says we must not climb Uluru because it is a spiritually significant place to the Anangu people of the Northern Territory.

Rules

Rules are sets of regulations or instructions. They are understood by people in a particular group or organisation. They tell you what you are allowed or not allowed to do. Rules exist to make sure that everyone acts in a way that is fair to others and in an organised way.

Most sports have a set of rules that apply just in that sport. In rugby league, there is a rule saying that the ball cannot be passed forward. In netball, you cannot make physical contact with another player.

Other organisations such as schools and libraries have rules. There are even rules for using punctuation and grammar when writing!

Differences between laws and rules

The main difference between laws and rules is what happens when they are broken.

When laws are broken, there are significant legal consequences. This may involve being punished by getting a warning from the police, paying a fine, paying someone else for the damage you caused, losing your right to participate in an activity (such as drive a car) or even going to prison.

When someone breaks the rules, the consequences aren't as severe as breaking a law. It might be that someone ends up being upset, you might have to fix something that was broken, you might temporarily lose a privilege, or you might have to serve a 'time out' penalty. These types of consequences for breaking rules allow the person to think about their actions and understand how their actions negatively impacted others before rejoining the activity.

Illustrator: Paul Lennon

TARGETING HASS 4 © PASCAL PRESS ISBN: 9781925726053

What is the difference between rules and laws and why are they important?

UNIT 29

Research

Choose a sport or activity that you do not participate in. What rules does it have? What are the consequences for breaking those rules?

Questioning

Rule or law?

1 No running in the classroom.

2 No driving a car without a licence.

3 Do not put your shoes on the lounge.

4 Wear a helmet when riding a bicycle.

5 Brush your teeth before going to bed.

6 You must attend school until you are 17.

Analysing

Rules are there to keep us safe. So, is it good to have lots of rules? Explain why.

Communicating

With a partner, discuss a rule that you think we don't need anymore.

Can you convince your partner to agree with you?

Evaluating and Reflecting

How do we know if a rule or law is a good one? Give some examples to explain your ideas.

UNIT 30

Plastic Bag Laws

Single use plastic bags, provided by retail shops to their customers, have become a significant environmental problem. Millions were used every year across the country, with many of them ending up as litter. Plastic bags choked our waterways and killed marine life such as turtles. Worst of all, each plastic bag took up to 1000 years to break down.

On 1 July 2018, the government of Western Australia enacted a law to ban single-use plastic bags. The law is called the *Environmental Protection (Plastic Bags) Regulations 2018*. Queensland enacted their ban at the same time, following on from South Australia, Tasmania, the Northern Territory and the Australian Capital Territory. Businesses and their customers were given six months to prepare for the change.

The law applies only to single-use plastic bags that are less than 35 microns in thickness. It does not include the plastic bags people use to package their fruit and vegetables. People have been encouraged to use fabric bags or thicker plastic bags that can be reused multiple times. The aim of the law is to reduce the number of plastic bags going to landfill and to minimise the environmental harm they cause.

The consequences for retailers who do not ban single-use plastic bags are severe. They will face a $5000 fine for each offence. Plastic bag suppliers who give misleading information about the thickness of their bags and the materials used to make them could also be fined $5000 or prosecuted in a court of law. Although the state government is in charge of enforcing the law, shoppers can report non-compliant retailers to the National Retail Association.

TARGETING HASS 4 © PASCAL PRESS ISBN: 9781925726053

Research

Find out about plastic bag laws in your own state. What are they? When did they begin? What are the penalties?

Questioning

Write four questions that can be answered from your Research.

What

When

Who

Why

Analysing

Do you think the ban on plastic bags will work? Will it solve the environmental problems caused by waste plastic? Explain your opinion.

Communicating

Create a poster to encourage people to carry their shopping without using plastic bags.

Be creative and imaginative!

Evaluating and Reflecting

Is the ban on single-use plastic bags a good law? Explain your reasons.

Sharing Culture: Uluru

Right in the middle of Australia stands a huge red rock – Uluru.

The rock and the country around it belong to a group of Aboriginal people called Pitjantjatjara and Yankunytjatjara. They are very difficult words. So, they say, call us Anangu (The people).

Anangu have lived here for tens of thousands of years. They are connected to their land, the plants and the animals through Tjukurpa. Tjukurpa is many things. It is the time of creation when heroic ancestral beings made the land. It is the law that governs and guides all the people's actions. It is the life force joining the people to the natural world. Ceremony and sacred places make the people one with Tjukurpa.

Because of Tjukurpa, Anangu know their lands so well that they can always find food and water, even during drought.

They lived well in good times. More importantly, they survived in the harshest years, even in the days before they owned cars or had a supermarket to shop at. Everything they needed for body and spirit they found in the land.

Anangu own the land. But they know that this place is important to other Australians as well, and they are very happy to share it with visitors from all over the world – more than 400 000 come every year. Anangu generously share much of their culture with all of these visitors. Tony Tjamiwa says, "Our purpose is to explain and clarify our understanding of our world so that others can understand."

An important part of Anangu law is that people should not climb the rock – Uluru. To Anangu it is a sacred place. In fact, on 25 October 2019, the climb up Uluru was permanently closed. Thirty-four years after the Anangu people were given back the title to the land, their desire to keep people off the rock will be enforced by law. Instead of climbing Uluru, Anangu urge visitors to go on the Mala (wallaby) and Liru (snake) walks, which are guided by Anangu and other rangers.

Through their parents, grandparents, aunts and uncles, Anangu children learn about Tjukurpa. They learn the songs and dances for inma (ceremonies). The girls are painted for the Kuniya (python) and the boys act the story of Lungkata, the blue-tongued lizard. The stories told come from the creative time. It is Anangu's duty to look after Tjukurpa. The elders say, "It provides us with our reason for living. Our law is how it should be." Inma are important; they teach young people to keep the Law properly in their hearts and minds.

Source: Sharing Culture: *Uluru*, pp. 1-2, p. 18, p. 53, Steve Parish

CIVICS AND CITIZENSHIP

TARGETING HASS 4 © PASCAL PRESS ISBN: 9781925726053

Research

What happened on the last day of climbing, before Uluru was permanently closed by law?

Questioning

What are the meanings of these Anangu words?

- Inma (in-ma)
- Tjukurpa (chook-orr-pa)
- Kuniya (koon-e-ya)
- Lungkata (loong-cart-ah)
- Anangu (a-na-nu)
- Mala (mahr-la)
- Liru (lear-oo)

Analysing

Based on your research, how did different people react to the closing of the climb up Uluru?

Communicating

What would you say to someone who still wants to climb Uluru?

Evaluating and Reflecting

Is it important to learn about the beliefs and laws of other groups of people? Explain why.

UNIT 32

Diwali

Hindus pray to Lakshmi, the goddess of wealth, during Diwali.

Diwali is a Hindu celebration, known as the Festival of Lights. Diwali means 'rows of lighted lamps'. The word comes from an ancient language of India called Sanskrit. In Fiji and Singapore, where many Hindus live, the celebration is a national holiday.

Diwali is celebrated all over the world. It runs for five days during late October or early November and marks the start of the Hindu New Year. The exact dates change each year because they are determined by the position of the moon. Hindu people worship Lord Ganesha for health and success and Goddess Lakshmi for wealth and wisdom.

People decorate floors with rangoli designs during Diwali.

Diwali sweets called jelebi, laddoo and unniappam are popular.

On the last day of Diwali, brothers visit their married sisters.

The festival is a celebration of hope. People wish for good luck. They remember stories of good beating evil. People decorate their homes and shops with lights and small clay lamps called diyas. It is a time for vibrant colour, watching fireworks and delicious food. The lights are symbols of hope.

During Diwali, people share gifts of sweets and dried fruit. They eat meals with friends and family. Many people clean their homes and put on new clothes. Hindu communities also give money to charities and do kind deeds for others.

Many towns and cities in Australia celebrate Diwali. Hindu representatives say that they want everyone in their community and beyond to enjoy the festivities. Everyone is welcome to join in.

Source: Go Facts Geography: *Special Days*, pp. 16–17, Blake Education

Research

Investigate Rangoli art, its designs, colours and materials.

Questioning

After reading the text, what questions do you have about the Hindu religion and Diwali?

Write at least four questions.

Analysing

Choose one of the groups that you belong to. Compare this group's celebrations with the Hindu celebration of Diwali.

How are they similar and different?

Communicating

Create your own Rangoli-style design as a symbol of hope and good luck.

Evaluating and Reflecting

How does being a member of a group impact a person's identity?

History Assessment 1

Why did the great journeys of exploration occur?

Task

- Select **one** great explorer. Complete the information below.

Who were they?	Why did they explore?

When and how did they travel?	What did they 'discover'?

- Show their journeys on the map below. Use a key to indicate different journeys.

- Analyse their journeys.

Positive	Negative	Interesting

ASSESSMENT

TARGETING HASS 4 © PASCAL PRESS ISBN: 9781925726053

History Assessment 2

Why did Europeans settle in Australia?

Task

- On the sails of the ship, write facts that you have learned about the First Fleet.

- In the space below, or using a digital device, create a flyer or poster encouraging soldiers and sailors to sign up for the voyage to the colony of New South Wales.

Illustrator: Paul Lennon

History Assessment 3

What was the nature and consequence of contact between Aboriginal and Torres Strait Islander Peoples and early traders, explorers and settlers?

Task

- What does this Aboriginal painting tell you about their contact with early explorers and settlers?

Aboriginals painted what they saw – including visitors to their land.

- What do these two images tell you about the consequences of the contact between Aboriginal and Torres Strait Islander peoples and early explorers and settlers?

Bennelong often wore clothes the British gave him.

An Aboriginal man from Botany Bay

Source: Go Facts Australia: *First Contacts*, p. 5, p. 9, p. 13, p. 17, Blake Education; *Blake's Australian History Guide*, p. 25, Pascal Press

ASSESSMENT

Geography Assessment 1

How does the environment support the lives of people and other living things?

Task

Select **one** country from the continent of South America. On the map, indicate:

- the national borders
- mountain ranges
- river systems
- the capital city
- significant natural features
- significant human features.

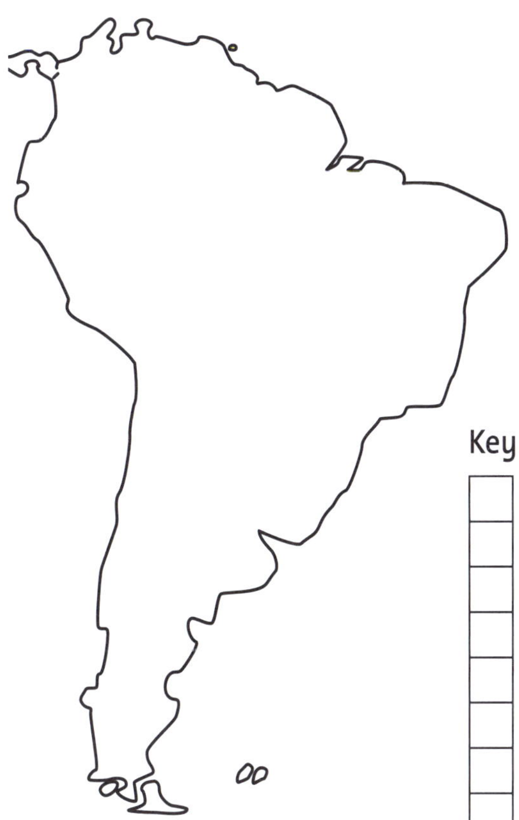

What environmental issues are faced by the country?

____

____

Of these issues, which one do you think is most important? ____________________

Why is it so important?

____

What caused this issue?

____

What can be done to try and solve this issue?

____

____

Geography Assessment 2

How can people use places and environments more sustainably?

Task

Conduct a playground survey about waste.

- At the beginning of lunch, survey the students in your class or grade and collect data on the types of food packaging they have.
- At the end of lunch, survey the same students to find out what they did with their food packaging.

Before Lunch

Types of packaging	Plastic film	Alfoil	Paper bag	Hard plastic container	Fresh fruit, no packaging	Soft plastic pouch, e.g. yoghurt	Plastic bag	Resealable plastic bag	Other
Number									

After Lunch

What students did	Took home to reuse	Put in garbage bin	Put in recycling bin	Threw on ground	Put in compost bin	Can't remember
Number						

At the end of lunch, look around your playground.

- How many garbage bins are available for students to use? __________
- How much litter is on the ground? ______________________________
- What impact does food packaging have on your playground?

____

- How can you and your school improve the waste situation in the playground?

____

- Write a letter to your school principal outlining your concerns about food packaging at school and making some suggestions for change.

Dear ______________________________,

____

____

____

____

____

____

ASSESSMENT

TARGETING HASS 4 © PASCAL PRESS ISBN: 9781925726053

Geography Assessment 3

How do different views about the environment influence approaches to sustainability?

Task

1 Using an atlas or website, on the map below identify the world's coral reef systems.

2 Describe the location of coral reefs.

____

____

3 Write the names of four different countries with waters containing coral reefs.

__________ __________ __________ __________

4 What is the climate like in areas that feature coral reefs?

____

5 What environmental threats do coral reefs face?

____

____

6 How can our approaches to sustainability impact these threats to coral reefs?

____

____

____

Civics and Citizenship Assessment 1

How can local government contribute to community life?

Task

- Look at the photograph of the playground.
- Write a letter to your local council. Tell them what they need to do to improve the park and explain why this is their responsibility.

Image by Trevor Marron 2018 Image cc attribution https://creativecommons.org/licenses/by/2.0/

Dear ______________________________,

ASSESSMENT

TARGETING HASS 4 © PASCAL PRESS ISBN: 9781925726053

Civics and Citizenship Assessment 2

What is the difference between rules and laws and why are they important?

Task

1 Identify one important law.

How would your community be different if this law did not exist?

2 Identify one important school rule.

How would your school be different if this rule did not exist?

3 Identify one important sporting rule.

How would the game be different if this rule did not exist?

4 Identify one important rule you have at home.

How would your family life be different if this rule did not exist?

5 If you were in charge of your sporting club, school or other organisation for a day, what new rule would you bring in, and why?

Answers

Note: Answers are not supplied to open-ended questions, where students' responses will vary.

ANSWERS

1 A Wider World

Questioning:

heathen: a person who does not belong to the religion followed by others
globalisation: the way a business or organisation develops influence in other countries across the world
kraken: a large mythical sea monster
diphtheria: a contagious disease
uncharted: an area of land or sea not mapped

Analysing:

Risks faced by early sea explorers	Risks faced by space explorers
Travelling through unknown territory with no or unreliable maps Contracting new diseases Not enough food and water Risk of injury without medical help	Travelling through unknown territory Loss of oxygen Availability of food and water No medical help nearby

2 Trade during the Age of Discovery

Analysing: During the Age of Discovery, Europeans did not know of the existence of Australia, so there was no trade between them. The trading centres from Europe, Africa and America are all on the coast so that they could be reached easily by ship.

Communicating: Example responses include: having access to different goods and products, creating jobs, providing more choice in what people buy, buying goods when they are not in season in one country.

Evaluating and reflecting: Answers will vary, but may include: we now grow/mine/produce some of these products in Australia so we do not have to trade and buy them from other countries. The trade in tobacco has changed because people do not smoke as much since we now know of the health issues around smoking. The fur trade is also less because of the changes made by animal rights organisations and people wearing different types of clothes now.

3 Exploration in the Past

Questioning:

1. He disproved the myth that the world was flat.
2. Columbus thought that he had sailed to India, but he was actually in America – a new continent.
3. Magellan had to find a better route to the Spice Islands.
4. The two important achievements of Magellan's journey were being the first expedition to sail from the Atlantic Ocean into the Pacific Ocean, and being the first expedition to circumnavigate the globe.

Analysing: Answers will vary, but may include references to the fact that America was already inhabited by many different groups of people (American Indian tribes, for example) and so Columbus could not have 'discovered' it.

5 Torres Strait Islander Peoples

Research: The green panel represents the land, the blue represents the water of the Torres Strait, the black stripes represent the people, the five-pointed star represents the five major island groups in the strait, the centre symbol represents their headdress (Dhari), the white represents peace.

Questioning:

1. The Torres Strait Islander peoples have Melanesian origins, they are not descendants of mainland Aboriginal peoples.
2. Sahul was the landmass that linked Southern New Guinea to Cape York Peninsula in Queensland.
3. Outriggers are the long weights used on the sides of the canoe to help it stay upright.
4. The Arafura Sea and Coral Sea

Analysing:

Similarities	Differences
Both groups have been in Australia for tens of thousands of years. Both groups were hunter/gatherers and lived off the land/sea. Both groups had Dreamtime/creation stories.	Torres Strait Islanders have Melanesian backgrounds. Torres Strait Islanders travel by canoe. Torres Strait Islanders were more likely to trade with other people.

Evaluating and reflecting: Answers will vary, but may include references to land areas being covered by water, cutting off land access to the continent, changing the way people moved and the access they had to food/game. If the land bridge had not been covered, the development of Aboriginal culture may have been different. The plants and animals that are unique to Australia may not be the same.

6 First peoples: Tasmania and Western Australia

Research:

- Birak (first summer) season of the young which is dry and hot
- Bunuru (second summer) season of adolescence which is the hottest part of the year
- Djeran (Autumn) season of adulthood when cooler weather begins
- Makuru (winter) season of fertility which is the coldest and wettest time of year
- Djilba (first spring) season of conception which has cold nights and warm days

TARGETING HASS 4 © PASCAL PRESS ISBN: 9781925726053

Answers

Note: Answers are not supplied to open-ended questions, where students' responses will vary.

• Kambarang (second spring) season of birth with longer dry periods

Analysing: Answers will vary, but may refer to their outdoor lifestyle, close ties to the natural world, no need for lots of clothing, use of spears for hunting, cooking food around a campfire, the importance of family groups.

7 Aboriginal Shell Middens

Research: Erosion is the action of wind, water or other natural agents to wear away the surface of rocks and soil. It impacts sites by removing parts of the site so that they are lost or destroyed. But erosion can also remove material that has covered sites so that they can be seen for the first time in hundreds or thousands of years, allowing people to learn more about them.

Questioning: Answers will vary, but may include references to what people ate, what sorts of tools they used, and how many people were in the local area.

Analysing: The sites would vary by what was found in them. They would include items that were local to a particular area. They would vary in how much material was in them, depending on the number of Aboriginal people who lived in the area.

8 The British are Coming

Analysing: Answers will vary, but may include:

Life in the 18th and 19th centuries	Life now
Lots of homeless children Children went to work in factories. Children didn't have to go to school. Children could go to jail.	Children have to attend school. Children have lots of play time. Access to computers and technology Children don't go to jail.

9 Arrival of the First Fleet

Research: Answers will vary. The fleet left England and went to Rio de Janeiro, then to Cape Town and then across the Great Southern Ocean to Botany Bay. The fleet experienced bad weather and wild storms. Food supplies were limited, so the convicts were almost starving.

Questioning:

1. Surgeon White was impressed by the size of the harbour.
2. BC meant the baby was born in the colony.
3. Many of the buildings collapsed because most of the convicts didn't know how to build, they were built too quickly, and the convicts used materials that didn't withstand the summer storms.
4. Refusing to work

Analysing: The main factors were the weather, the different seasons because they were in a different hemisphere, the lack of skilled people, the lack of food supplies.

11 Food Rationing

Research: Answers will vary, but should include a statement about seasons.

Questioning:

rationing: the control of a scare substance such as food
scurvy: a disease caused by the lack of vitamin C
rancid: old and stale food that smells or tastes unpleasant
mutton: meat from a sheep that is more than 2 years old
self-sufficient: do not need help from others, can produce food for yourself

Analysing: They were unable to plant crops easily and had difficulty digging up large areas of ground for farming. It made food production more difficult.

12 The Macassans

Research: Answers will vary. Example response: The Yolngu are Aboriginal people who live in north-eastern Arnhem land. Children belong to the same moiety (division) as their father. These divisions are either Dhuwa or Yirritja and everything in Yolngu country (plants, animals, spirits, land, water) belong to one of these two divisions. There are different clans within each moiety which has its own language.

Questioning: 1. false, 2. true, 3. false, 4. true, 5. false, 6. true

Analysing: Answers will vary, but may refer to it being a mutually beneficial relationship, where both peoples got items they needed.

Communicating: Answers will vary, but may refer to cooperation, use of tools and different types of shelter.

Evaluating and reflecting: Answers will vary, but may include reference to greater levels of cooperation and less violence and dispossession.

13 Assistance, Resistance and Reprisal

Research: Pemulwuy resisted the movement of European settlers into his people's traditional lands. Pemulwuy was upset that the settlers were breaking Aboriginal law, taking away their rights and he opposed their violence towards traditional Aboriginal landowners. He stole from them. He raided their farms. He was involved in the death of John McIntyre, who was Governor Phillip's gamekeeper. At first, Governor Phillip had been told to be polite and friendly towards the "Indians" (as they called the Aboriginal people). But when McIntyre died, Governor Phillip ordered a reprisal. The next Governor, King, ordered that Aborigines near Parramatta could be shot at sight and Pemulwuy was declared an outlaw.

Analysing: Bennelong lost the respect of his tribe because he wore European clothing and lived and acted like one of them.

Answers

Note: Answers are not supplied to open-ended questions, where students' responses will vary.

ANSWERS

14 Truganini

Research: Answers will vary, but will include the fact that Truganini joined with four other Tasmanian Aboriginals and they robbed and shot at settlers around Dandenong, Bass River and Cape Paterson. They were chased by the authorities. Some members of the group killed two men who were whalers.

Analysing: They were called the 'Black wars' because that is how they were seen by the European settlers. It made the conflict about race and colour, rather than the actions that caused it. A more appropriate name could refer to the location or time of the conflict.

15 Weather and Climate Around the World

Research:

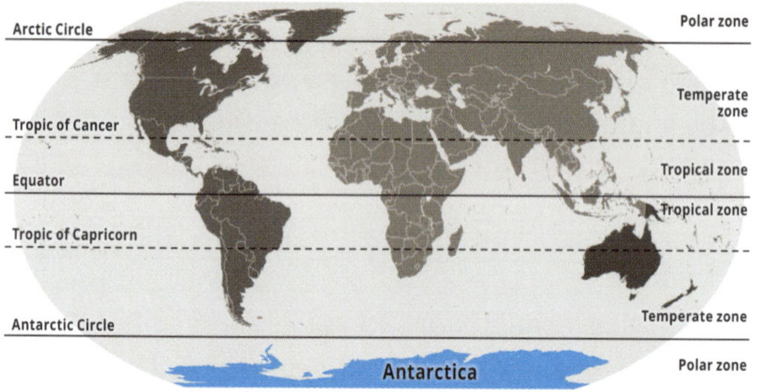

The torrid zone is the area between the Tropic of Cancer and the Tropic of Capricorn. Latitude 23.5 degrees north to 23.5 degrees south.

Questioning:

1. The Tropic of Capricorn
2. They are both dry and have low levels of precipitation.
3. Tropical or torrid zone
4. in deserts
5. New Zealand would be cool temperate. It would have some snow in winter, regular rainfall but usually mild conditions.

Analysing: As you move away from the equator, the climate becomes cooler and there are fewer violent storms. This is because the equator is the part of the earth that is closest to the sun.

16 South America

Research: An isthmus is a narrow strip of land with sea on either side of it that links two larger areas of land. They are found all over the world. Some examples are: Gibraltar, Panama, Suez, and, in Australia, Barrenjoy headland and Eaglehawk Neck.

Analysing: Answers will vary, but may refer to the similar land/ soil conditions.

17 Africa

Analysing: The stinging thorn provides the ants with a safe home and nectar to eat. The ants provide the stinging thorn with protection from predators that would eat the plant.

18 Australia's Living Resources

Research: Possible answers include:

Woodland: Has an open canopy. Sunlight reaches the forest floor so there is limited shade and moisture. They can be transition zones between different ecosystems. They are low density forests. They have some shrubs and grasses.

Swampland: Land that is permanently filled or covered in water. They exist everywhere except Antarctica. They occur near or along large rivers and lakes. They can be very fragile.

Coastal plain: Flat, low-lying land next to the ocean (coastline). Its elevation (height) does not vary a lot. They feature low-level shrubs and any trees are short or stunted.

Forest: Very similar to woodlands. They are large areas of land covered with trees and undergrowth. There are many different types of forest depending on the type of climate and level of rainfall, from temperate forests to rainforests.

Questioning: 1. true, 2. true, 3. false, 4. false, 5. true, 6. false, 7. false, 8. false

Analysing: So there is little waste and so the plants can be used sustainably.

19 The Land Affects Us

Analysing: Because it is too dry. People need water to survive: to drink, grow crops, water animals that help us and give us food. Without water, we cannot survive in an area for very long.

20 Guarding Ecosystems

Questioning:

1. An ecosystem is an interconnected community of living and non-living things that rely on each other.
2. The numbers of orang-utans have fallen in Malaysia due to forests being cleared so that palm oil can be grown.
3. Forest clearing lowers ecosystem diversity.
4. Orang-utans eat the plants and spread the seeds through the ecosystem.
5. Palm oil plantations are found in Malaysia, Borneo and Indonesia.

Analysing: The purpose is to create an emotive reaction from the reader. Readers feel upset and sad for the loss of habitat for the orang-utans.

TARGETING HASS 4 © PASCAL PRESS ISBN: 9781925726053

Answers

Note: Answers are not supplied to open-ended questions, where students' responses will vary.

Communicating: Answers will vary, but here is a sample answer:
hunt wolves – more elk – less willow trees – less birds.

21 Extreme and Unique Environments

Research: Only found in South Australia in the dry salt lakes. It has adapted to cope with high temperatures and lack of water. The lizard burrows into the damp sediments under the salt crust. It uses body movement and postures to maintain the right body temperature. Its skin doesn't let water evaporate from its body quickly.

Questioning:
1. They are very specialised so small changes can have a big impact.
2. Salt is produced by weathering rock or is left over from when the land was covered in salt water, or salt that has fallen in rain.
3. The Mountain Pygmy-possum lives in the alpine areas of Mt Buller, Mt Hotham and Mt Kosciuszko.
4. The North Sydney harbour program protects a colony of Little Penguins.
5. A niche habitat is one where an animal or plant has adapted to suit a very particular environment and its features.

Analysing: Plants and animals have little hope of surviving anywhere else because they have adapted to a very small range of environmental features and conditions.

22 Ancient Land Management Practices

Research:
Fish: sharks (great white, tiger, whale shark), golden perch.
Birds: curlew, sharptailed sandpiper, whimbrel, godwit, red-necked stint, plover, snipe
Answers will vary, depending on the animals selected.

Questioning: 1. false, 2. true, 3. false, 4. true, 5. false, 6. true

Analysing: They don't overuse the resources in one particular area. They make sure they have some left to grow for the next season. They use all parts of each plant and animal, so nothing is wasted. They move to different areas to use what is plentiful during different seasons. They encourage new growth through their farming methods.

23 Connecting to Land

Analysing: Their spiritual connection means that it is their duty to look after the land, to care for it and not use it unwisely or cause damage to it. They want to make sure that the land and its plants and animals are there for the next generation.

25 Energy from the Earth

Analysing: Other countries can learn from what Thisted has done. They can take the same actions in their own countries to improve their energy production and use. Thisted has made a positive contribution to improving the environment and lowering the use of non-renewable resources.

26 Living Sustainably to Conserve Wildlife

Research: Ecotourism is a type of tourism where visitors are encouraged to visit places because they are natural, undisturbed by human actions, have sustainable practices and are small in scale to limit the environmental impact. It is important because it creates options for people and it helps to promote sustainability. The benefits are lowering the impact of tourism on a specific place and the plants and animals that live there, and increasing people's appreciation of nature.

Questioning: 1. fact, 2. opinion, 3. opinion, 4. opinion, 5. fact, 6. fact

Analysing: Australia has a large ecological footprint because we have a lot of land and because we use a lot of natural resources. Mining and farming are two very big industries in Australia.

27 Understanding your Local Council

Analysing: Answers will vary, but should involve dot points from the stimulus text.

28 Local Council Case Study

Questioning:
- People who want to be more active, people who do not want to drive a car to work or school, people who want to visit the area.
- They will benefit by having a safe way to travel that is not on the road – it is just for bikes.
- State and local governments are involved.
- They want to attract tourists who are physically active and likes sports.
- Local residents and business owners who might have to give up land for the bike paths might be disadvantaged. Also, people who use areas near the paths will be disadvantaged during construction due to noise, pollution and restricted movement.

29 Laws and Rules

Questioning: 1. rule, 2. law, 3. rule, 4. law, 5. rule, 6. law

31 Sharing Culture: Uluru

Research: On the last day, there was a rush of people who lined up to climb Uluru.

Questioning: Inma – ceremonies; Tjukurpa – the law/the land; Kuniya – python; Lungkata – blue-tongued lizard; Anangu – people; Mala – rufous hare-wallaby; liru – king brown snake

32 Diwali

Research: Rangoli is an art form from India. Rangoli designs are created on the floor or the ground and use materials such as coloured rice, flour, sand or flower petals. Rangoli decorations are supposed to bring good luck. Designs are geometrical, or may represent creatures.

Answers

Note: Answers are not supplied to open-ended questions, where students' responses will vary.

ANSWERS

History Assessment 1

Sample responses

Christopher Columbus

- To find new trade routes to India, China and Japan.
- 1492, 1493, 1498 & 1502, on sailing ships.
- Claimed the Bahamas for Spain, explored islands in the Caribbean Ocean, travelled along the eastern coast of Central America.

Vasco da Gama

- Portugal wanted a maritime trade route to India.
- 1497-1499, 1500-1502, 1524-1524, on sailing ships.
- Linked Europe and Asia by finding a sea route to India.

Ferdinand Magellan

- Wanted to find a western route to the Spice Islands and create a useful trade route for Spain.
- 1519-1521, on sailing ships.
- First circumnavigation of the Earth.

Luis Vaz de Torres

- Searching for the continent Terra Australis (Australia).
- 1606-1607, on sailing ships.
- Claimed New Guinea for Spain, proved that New Guinea was not part of a larger continent (Terra Australis).

Willem Janszoon

- Wanted to find routes and markets for trade.
- 1605-1606, on sailing ships.
- First European known to have seen the coast of Australia, made landfall on Cape York, Queensland.

Zheng He

- Developing trade in Indonesia, Thailand and South East Asia as well as Africa, and expanding China's power and influence.
- 1405-1424, on sailing ships.
- Followed previously established trade routes.

La Perouse

- Leading an expedition around the world – scientific exploration.
- 1785-1788, on sailing ships.
- First European to set foot on the island of Maui (Hawaii), encountered the First Fleet off Botany Bay, 24 January 1788.

Abel Tasman

- To find out about a suggested large southern land mass.
- 1642-1643, 1644, on sailing ships.
- Discovered Tasmania and named it Van Diemen's Land, the First European to sight New Zealand.

History Assessment 2

Sample responses:

- 11 ships
- left Great Britain on 13 May 1787
- transported convicts
- commanded by Captain Arthur Phillip
- journey lasted 8 months
- 2 navy ships (*HMS Sirius, HMS Supply*)
- 6 convict transports (*Alexander, Charlotte, Friendship, Lady Penrhyn, Prince Of Wales* and *Scarborough*)
- 3 supply ships (*Golden Grove, Fishburn* and *Borrowdale*)
- stopped in Santa Cruz, Rio de Janeiro and Cape Town
- some ships arrived in Botany Bay on 18 January 1788, the others two days later
- met in Botany Bay by the local Cadigal people

History Assessment 3

Answers will vary. They could include comments about seeing sailing ships for the first time, different types of animals, and the introduction of metal tools such as knives and axes.

Answers will vary. They could include comments about making Aboriginal people dress and act like Europeans, changing their beliefs, making Aboriginal people less traditional.

Geography Assessment 1

Sample responses:

Brazil

Capital: Brasilia
Rivers: Amazon, Parana, Rio Negro
Mountain ranges: Mantiqueria mountains, Espinhaco Mountains, Serra do Mar
Natural features: Sugarloaf Mountain, The meeting of the waters, Cachoeira Fumaca, Iguazu Falls
Human features: Cristo Redentor, Palacia do Itamaraty, Carnaval
Environmental issues: deforestation, illegal wildlife trade, water pollution

Argentina

Capital: Buenos Aires
Rivers: Parana, Uruguay River, Paraguay River, Iguazu River
Mountain ranges: Andes, Sierras de Cordoba, Martial Mountains
Natural features: Iguazu Falls, The Ibera Wetlands, Valdes Peninsula, Perito Moreno Glacier
Human features: Palacio Barolo, Kavanagh Building, Casa Curutchet, Cerro Uritorco
Environmental issues: pollution, soil erosion, soil salinisation, loss of agricultural land

TARGETING HASS 4 © PASCAL PRESS ISBN: 9781925726053

Answers

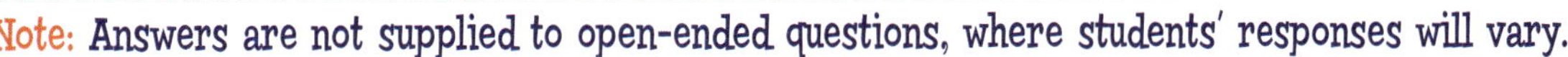

Note: Answers are not supplied to open-ended questions, where students' responses will vary.

Peru
Capital: Lima
Rivers: Amazon, Maranon River, Ucayali River, Huallaga River
Mountain ranges: Andes, Cordillera Blanca, Cordillera Huayhuash
Natural features: Ballestas Islands, Paracas Desert, Lake Titicaca, The sacred valet of the Incas
Human features: Machu Picchu, The Inca Trail, ruins of Cuzco
Environmental issues: loss of rainforest, deforestation, water pollution, soil erosion

Chile
Capital: Santiago
Rivers: Maipo River, Biobio River, Futaleufu River
Mountain ranges: Andes, Chilean coast range, Cordillera Domeyko
Natural features: Atacama Desert, Glaciers of Tierra del Fuego, granite walls of Torres del Paine
Human features: Chiza geoglyphs, Geoglyphs of Pintados
Environmental issues: deforestation, soil erosion, pollution

Venezuela
Capital: Caracas
Rivers: Orinoco, Rio Negro, Rio Casiquiare
Mountain ranges: Andes, Guiana Shield, Cordillera de Merida
Natural features: Angel Falls, Orinioco Delta, Mochima National Park
Human features: National Pantheon of Venezuela, Statue of Virgin Mary in Trujillo
Environmental issues: sewerage pollution, deforestation, soil degradation

Ecuador
Capital: Quito
Rivers: Rio Naop, Rio Pastaza, Rio Cararay
Mountain ranges: Andes, Cordillera Real, Cordillera del Condor
Natural features: Galapagos Islands, Amazon Basin, El Cajas National park
Human features: Cojitambo Canar Inca Ruins, Santa Anna, Loma Alta marble figurines
Environmental issues: land erosion, deforestation, flooding, desertification

Geography Assessment 2

Students' responses should show an awareness that it is important to reduce our rubbish because we have limited resources and it can be expensive to recycle some products. Their results should show if people are recycling the items that can be recycled, and if people are relying too much on materials that are difficult to recycle or reuse.

Ideally, students will follow up with the school organisation about making changes to food packaging.

Geography Assessment 3

2. Coral reefs are located in shallower waters closer to the equator.
3. Answers will vary, but include Australia, Papua New Guinea, Indonesia, Malaysia, Philippines, Madagascar, India.
4. The climate in areas with coral reefs is warm and tropical. There is lots of sunlight penetration of the water.
5. Environmental threats to coral reefs are coral bleaching, sediment, crown of thorns starfish, pollution.
6. Our approaches to sustainability can limit the rises in ocean water temperature, we can reduce sediment run-off through better agricultural techniques, and we can reduce plastic pollution.

Civic and Citizenship Assessment 1

Answers will vary, but may include references to rusty and broken equipment; the need to care better for the garden/grass/park area; the need to add shade, fencing and footpaths; the need to provide garbage/recycling bins; and the need to add softfall.

Civic and Citizenship Assessment 2

Answers will vary.

Targeting HASS
Year 4

Copyright © 2020 Pascal Press
ISBN: 978 1 925726 05 3
Reprinted 2022, 2025

Published by Pascal Press
PO Box 250
Glebe NSW 2037
contact@pascalpress.com.au

Author: Merryn Whitfield
Publisher: Lynn Dickinson
Typesetter: Ruth Schultz
Editor: Ruth Schultz
Designer: Janice Bowles

Printed by Vivar Printing/Green Giant Press